Fanfare for Earth

Also by Harry Y. McSween, Jr.

Stardust to Planets: A Geological Tour of the Solar System

Meteorites and Their Parent Planets

Geochemistry: Pathways and Processes

Fanfare for Earth

The Origin of Our Planet and Life

Harry Y. McSween, Jr.

St. Martin's Press
New York

Production Editor: David Stanford Burr

Library of Congress Cataloging-in-Publication Data

McSween, Harry Y., Jr.
Fanfare for Earth : the origin of our planet and life / by Harry Y. McSween, Jr.
p. cm.
ISBN 0-312-14601-9
1. Earth—Origin. 2. Life—Origin. I. Title.
QB632.M42 1996
551.7—dc20 96-23898
CIP

First Edition: March 1997

10 9 8 7 6 5 4 3 2 1

This book is dedicated to the children in my life

Cole, Fran, Gray, Harrison,
Justin, Kelly, Lindsay, Sydney,
and especially Pierce

Contents

Acknowledgments

By necessity, a book on the origin of our planet and life covers a lot of scientific real estate, and various chapters have touched upon subject areas with which I had only a passing familiarity. I thank my friends and colleagues Tom Broadhead, Bill Dunne, Mike Green, Ralph Harvey, Andy Knoll, David Kring, Mike McKinney, Fritz Schilling, Ross Taylor, and Edward Wing for enlightening discussions and for steering me to appropriate literature. For inspiration I am also indebted to the broader community of scientists, whose stimulating discoveries inform and amaze us all.

I appreciate encouragement and professional guidance from my New York connections, literary agent Julian Bach and St. Martin's science editor Keith Kahla.

And finally, I am grateful to my family for understanding this strange compulsion to write and for not begrudging the many hours I spend doing it.

How do you say to a child in the night,
Nothing's all black, but then nothing's all white . . .

Beginning lines of "Children Will Listen,"
from the Broadway musical *Into the Woods*
with permission of Warner Bros. Publications

Fanfare for Earth

Prologue

Sir James Jeans (1877–1946), lecturer in mathematics at Cambridge University in England, was a gifted scientist whose research was devoted to the physics of the very small (individual molecules of gas) and the very large (planets and stars). In 1928 Jeans was knighted, deservedly, for his seminal scientific contributions and for his service to the Royal Society. The next year he abruptly shifted gears to begin a second career and, for the remainder of his life, he was perhaps the world's leading spokesperson for popular science. His beautifully crafted books were resoundingly successful with the public, and he was in constant demand as a lecturer and broadcaster.

One of the recurrent themes of Jeans's work was that scientific concepts are most clearly, most fully, and most naturally explained by mathematics. Accordingly, he concluded that the "great architect of the universe" must be a mathematician (some of Jeans's contemporaries suggested, somewhat facetiously, that he was creating God in his own image). The point Jeans strove to make, though, was a deadly serious one: the "designing and controlling power" of the universe was not interested in emotion, morality, or aesthetics, but only in order, at least as far as could be divined from the study of his works. This conviction, when coupled with Jeans's scientific studies of the structure and evolution of the immense universe, left little room for foolish sentiment about the supposed importance of living creatures, even those capable of conscious thought, or of the one tiny blue planet that such beings are known to inhabit. In Jeans's own words:

> What does life amount to? We have tumbled, as though through error, into a universe which by all the evidence was not intended for us. We cling to a fragment of a grain of sand until such time as the chill of death shall return us to primal matter. We strut for a tiny moment upon a tiny stage, well knowing that all our aspirations are doomed to ultimate failure and that everything we have achieved will perish with our race, leaving the Universe as though we had never existed. . . . The Universe is indifferent and even hostile to every kind of life.
>
> —James Jeans, *The Mysterious Universe*

These hard words, written more than half a century ago, may seem extreme, but they are not too different from the musings of many other science writers since his time. But did Jeans go too far? Is this belittling view of the Earth and its denizens really what modern science strives to teach our children? I think not, and I intend to introduce a loftier perspective in this book.

Jeans's sentences echo centuries of skirmishes pitting science against myth, philosophy, and religious pronouncements about the creation and proper place in the cosmos of our world and its inhabitants. We now recognize that the Earth, once thought to be fixed at the very center of the universe, actually occupies a mundane position around a minor star in the outskirts of a spiral galaxy like many thousands of others. The idea that our location is not special is embodied in the so-called Copernican Principle, named for the man who first argued convincingly that the Earth revolves around the Sun. So, if there is anything at all special about our planet, it must be the peculiar circumstances that allowed life to appear and to evolve into riotous diversity. This, too, is an old battleground, with victories claimed by both science and religion. Indeed, to this day, religious beliefs and philosophy clash with scientific theories about the origin of the Earth and especially of ourselves. As expressed by anthropologist Loren Eiseley:

> Century after century, humanity studies itself in the mirror of fashion, and ever the mirror gives back distortions, which for the moment impose themselves upon man's real image. In one period we believe ourselves governed by immutable laws; in the next, by chance. In one period angels hover over our birth; in the following time we are planetary waifs, the product of a meaningless and ever altering chemistry. We exchange halos in one era

> for fangs in another. Our religious and philosophical conceptions change so rapidly that the theological and moral exhortations of one decade become the wastepaper of the next epoch. The ideas for which millions yielded up their lives produce only bored yawns in a later generation.
>
> —Loren Eiseley, *The Unexpected Universe*

Philosophy, religion, and science provide powerful yet distinct (sometimes complementary, sometimes contradictory) perspectives on origins. It is not really my intent to try to reconcile these different approaches. At its heart, this is a science book, though its content raises serious philosophical and (for some) religious questions.

James Jeans disavowed any acquaintance with philosophy other than as an intruder, but he correctly recognized that scientists have a necessary role to play in this arena:

> It should be the business of philosophy to criticize science rather than the business of science to criticize philosophy; but in practice few philosophers become acquainted closely enough with the actual day-to-day tactics of scientific advance, and take even the grand strategy of that advance from its scientific exponents rather than from the content of the advance itself.
>
> —Quoted by E. A. Milne, *Sir James Jeans, A Biography*

I have few pretensions about my own qualifications as a philosopher. I will borrow liberally from others, however, in pointing out some of the deeper meanings of scientific discoveries about the origin of our planet and its life.

The reader should be forewarned that some discomfort, perhaps even frustration, may result from examining the acts of creation through the eyes of scientists. The scientist's method of problem solving is rather plodding, in that only those ideas that pass rigorous experimental or observational scrutiny are accepted. Reacting to the criticism that science takes away from the beauty and mystery of the stars, the renowned physicist Richard Feynman once countered that it does "no harm to the mystery to know a little about it." Some will, of course, take issue with this premise, preferring a universe crafted by forces cryptic and unfathomable. Even scientists are not immune to the allure of mystery, while at the same time working to solve it. "The more the universe seems comprehensible," physicist Steven Weinberg once wrote, "the more it also seems pointless." Well, I disagree profoundly with the notion that a plain and pointless world is the price of understanding. My *goal* is to introduce some of the scientific advances, both historic and modern, that have led to a revolutionary understanding of its creation. My *hope* is to accomplish that without dispelling the sense of wonder that quite rightfully attends such revelations.

In the following chapters we will explore the origin of a seemingly ordinary, but actually quite unusual place—the third planet outward from the Sun, a once-pummeled and still-smoldering ball of stone and metal, a sphere whose distilled rocky crust wanders restlessly over its surface, a shimmering blue world awash in water and teeming with carbon-based life. As we go along, the reader will become privy to a secret or, if not that, at least a fact seldom appreciated and too rarely commented upon, even by those whose stock and trade are the very scientific advances that have divulged it:

> It looks as though modern science has been holding back a revelation quite as momentous as Copernicus's: it is Earth, not sky, that is heavenly.
>
> —Vincent Cronin, *The View from Planet Earth*

We will return to the broader meaning and importance of this special place later, but for now let us begin to explore, from the scientist's perspective, the origin and evolution of our world, its cosmic neighborhood, and its inhabitants.

Genesis to Geology

A Brief History of Thoughts About Creation

THE MAGIC OF BEGINNING

Witnessing the birth of a child is sheer magic, or so it seems if it is your own. I will never forget being mesmerized by my newborn daughter, twitching impossibly tiny fingers and toes under the glare of a warming lamp, as she opened her eyes on the world and me for the very first time. Such experiences make indelible impressions that last a lifetime.

But what is magic about it? Babies are born all the time—we are awash in humankind. I suppose that birth seems like magic because it epitomizes a beginning, the sudden appearance of a new life that was not here before. And it remains miraculous even though we might understand the biology of conception and gestation.

Archaic man sought to understand the magic of birth and of other aspects of the natural world by studying the heavens. In ancient times, it was thought that an individual's fortune was predetermined by the relative positions of the planets at the instant of his or her birth, that the Earth itself was situated at the center of the universe, and that deities regularly sent messages to mankind encoded in celestial light. In a sense, we have been living in this age of astronomical myth until just yesterday (and, judging by the persistence of belief in astrology, some of us still live there). Only within the last few hundred years have we begun to progress from astrology to astrophysics, from alchemy to chemistry, from Genesis to geology. But the transition has always been difficult and frustrating, and so it continues to this day.

This progression from myth to science posed a serious philosophical challenge even for the very innovators who forged the change in viewpoint. Many groundbreaking scientific discoveries of past generations were colored by worldviews consonant with mythology. For example, Johannes Kepler, the mathematician who unraveled the true orbital motions of the planets, was motivated by a desire to reveal the musical "harmony of the spheres." He sought to maintain a thread of continuity with past philosophy and, in so doing, revealed how truly difficult it was (or, more accurately, is) to let go of what we ourselves have been taught. Even Isaac Newton, the very epitome of rigorous scientific thinking, was not immune, as indicated by the following description:

> Newton was not the first of the Age of Reason. He was the last of the magicians, the last of the Babylonians and Sumerians, the last great mind which looked out on the visible and intellectual world with the same eyes as those who began to build. . . . Why do I call him a magician? Because he looked on the whole universe and all that is in it as a riddle, as a secret which could be read by applying pure thought to certain evidence, certain mystic clues which God had laid about the world. . . . He believed that these clues were to be found partly in the evidence of the heavens and in the constitution of elements, but also partly in certain papers and traditions handed down by the brethren in an unbroken chain back to the original cryptic revelation in Babylonia.
>
> —John Maynard Keynes, *Newton Tercentenary Celebrations*

If such consummate scientists as Kepler and Newton had difficulty disentangling their understanding of the orbital gyrations and gravitational tugs of celestial bodies from astronomical myth, how much more unfathomable to them must have been the *origin* of the heavens. Modern scientists, though less influenced by myth, still ponder the question of origins. This is difficult work, teasing the limits of what can be learned. Even fundamental concepts like time and space have no meaning before the creation. Moving backward in time and space past the first moment of creation we encounter the unknowable, and we have always had a difficult time conceptualizing the void, a universe replaced by nothing. In *Alice in Wonderland,* the Red King asks Alice, "What do you see?" "Nothing," she replies, and the Red King comments with admiration, "My, what good eyes you have."

CREATION STORIES

Creation myths are mankind's attempts to deal with the unknowable. But most myths do not stop there. They go on to try to explain the knowable, those events and processes of creation that now are amenable to scientific inquiry.

Despite the remarkable diversity among our planet's cultures and religions, a few recurring themes appear again and again in their lore about the origin of our world and its animal and vegetable denizens.

> In the beginning, one of the gods of heaven stirred the chaotic waters with his staff. When he raised the staff, muddy foam dripped down from it, this expanded and thickened, until it formed the islands of Japan. . . .
>
> —JAPAN

> Long, long ago there wasn't any land at all, only the ocean, but there was a god named Lowa who came down to an island. This god made a command followed by a magical sound, "Mmmmmm," and all of the islands were created. He went back to heaven and sent down men to this island. Each man had duties to perform—producing all living things, looking after the winds, and taking care of all death.
>
> After these men were in their places, Lowa sent another man down to arrange the islands. He put all the islands in a basket and, starting from the Carolines, put them into their present positions.
>
> Next Lowa sent two men down to tattoo everything in the world that had been created. This is how each kind of animal got its characteristic markings, and it also began the rank-signifying tattoos for chiefs, commoners, and women.
>
> —MARSHALL ISLANDS

> In the beginning, in the dark, there was nothing but water. And Bumba was alone.
>
> One day Bumba was in terrible pain. He retched and strained and vomited up the sun. After that light spread over everything. Bumba vomited up the moon and then the stars, and after that the night had its own light also.
>
> Still Bumba was in pain. He strained again and nine living creatures came forth: the leopard, the crested eagle, the crocodile,

and one little fish; next, the tortoise, and Tsetse, the lightning, swift, deadly, beautiful; then the white heron, also one beetle, and the goat. Last of all came forth men.

The creatures themselves then created all the other creatures. Then the three sons of Bumba said they would finish the world. The first, Nyonye Ngana, made the white ants; but he was not equal to the task, and died of it. Chonganda, the second son, brought forth a marvelous living plant from which all the trees and grasses and flowers in the world have sprung. The third son, Chedi Bumba, wanted something different, but for all his trying made only the kite bird.

When at last the work of creation was finished, Bumba walked through the peaceful villages and said to the people, "Behold these wonders. They belong to you."

—AFRICA, Bushongo

Once the earth was a tiny disk, only big enough for God and the Devil to lie upon together, and everything else was sea. The Devil did not care for this at all; he wanted to get rid of God and have the earth to himself. So he said to God, "You look tired. Why don't you lie down and sleep and I will keep watch?" God knew everything; he knew the Devil's mind precisely. Nevertheless he lay down and pretended to sleep. Then the Devil crept up behind him, got hold of God's shoulders and tried to push him into the sea. But he could not; though he seemed to be on the brink of it, the shore kept stretching on and on, the sea marched ever farther away. That was the north. So then the Devil got hold of God's shoulders and pushed him toward the south, and again it was in vain, again the shore expanded and try as he might he could not reach the sea. He tried pushing God to the east, the same; he tried pushing him to the west, the same again. Afterward the earth reached as far as the eye could see, and that was the beginning of the world.

—BULGARIA

A teal came flying, searching for somewhere to make herself a nest. She flew to west and east, north and south, saw nothing but sea, nowhere to settle. The Mother of Water took pity on her wanderings, she rose from the waves and crooked her knees to make a lap. The teal saw it thankfully as a peaceful green island.

She laid six golden eggs in the nest she made, a seventh of iron. The eggs rolled off one by one, fell down into the sea and shattered into fragments.

Yet no part of them was wasted, golden or iron—not yolk nor white nor shell—all were beautiful in making. Half of one eggshell formed the earth itself, its upper half arched the heavens over it. Its yolk made the golden sun, its white became the moon, its specklings were stars, its darker parts the clouds.

And still the Mother of Water lay rocking on the ocean. She rested for nine years; in the tenth she raised her head and began the business of creation. She put out one hand and there were all the headlands. She dipped her feet in the ocean and so made caves for fishes. She dived deep beneath the sea, with both hands formed its bed. She turned her head toward the land and extended the level shores, she made beaches and bays along it and good places for fishing. She planted rocks and reefs at sea to await unwary sailors.

—FINLAND

At first there was no earth and sky; there were only two great eggs. But they were not ordinary eggs, for they were soft and shone like gold. At last, as they went round they collided, and both the eggs broke open. From one came the earth, from the other the sky, her husband.

Now the earth was too big for the sky to hold in his arms and he said, "Though you are my wife, you are greater than I and I cannot take you. Make yourself smaller."

The earth accordingly made herself pliable and the mountains and valleys were formed, and she became small and the sky was able to go to her in love. When the sky made love to the earth, every kind of tree and grass and all living creatures came into being.

—INDIA

For a long period Ta'aroa dwelt in his shell. It was round like an egg and revolved in space in continuous darkness. But at last Ta'aroa was flipping his shell and it cracked, and broke open. Then he slipped out and stood upon the shell, and cried out, "Who is there?" No voice answered.

Ta'aroa overturned his shell and raised it up to form a dome for the sky. Then he took his spine for a mountain range, his ribs for mountain slopes, his arms and legs for strength for the earth; his fingernails and toenails for scales and shells for fishes; his feathers for trees and shrubs to clothe the earth; and his intestines for lobsters, shrimps, and eels. And the blood of Ta'aroa drifted away for redness for the sky and for rainbows.

But Ta'aroa's head remained sacred to himself, and he still lived. He was master of everything, and there was expansion and there was growth.

—TAHITI

Only when the God Phan Ku died was the world at last complete. The dome of the sky was made from Phan Ku's skull. Soil was formed from his body. Rocks were made from his bones; rivers and seas, from his blood. All of plant life came from Phan Ku's hair. Thunder and lightning are the sound of his voice. The wind and clouds are his breath. Rain was made from his sweat. And from the fleas that lived in the hair covering him came all of humankind.

—CHINA

MAKER OF HEAVEN AND EARTH

It is comfortable to suppose that the creation stories recounted above are relics of the dim past or of primitive cultures, and that sophisticated, modern people no longer need such mythical imagery. However, any modern science book impertinent enough to address topics like the origins of the Earth, its cosmic surroundings, and its scurrying and leafy inhabitants must eventually run smack into the brick wall of Genesis, which serves as the historical foundation for orthodox Jewish and Christian thought about creation. (The Islamic Koran does not contain a continuous story of creation, but various passages, taken together, reveal a common heritage with the Genesis account.) A recent survey in a popular magazine indicated that more than half of the respondents (all American adults) believe that Genesis provides an accurate account of how we came to be. Rather than blissfully ignoring this formidable biblical wall, I will face it directly, by discussing its history and its rightful place among mankind's noblest attempts to understand ourselves and our surroundings. My intent in so doing is not to demean anyone's religious beliefs, nor do I expect to change anyone's convictions about this issue. Nevertheless, the biblical account stands in stark contrast to the picture of creation that emerges from modern scientific research. To pretend (by ignoring) that there is not a widely held view that posits science is wrong about creation is folly. So as to gain some perspective on the nature of the disparity between the views of religion and science, let us first see exactly what the Bible has to say.

Genesis is a Greek word meaning "coming into being." Its use as the title of the first book of the Bible dates from the third century B.C., when the Bible was first translated from Hebrew into Greek. Genesis provides the biblical account of the creation of the world, but even a cursory reading indicates something perplexing: it contains not one, but two rather distinct accounts of creation. In fact, there are enough differences between these accounts that they must have had different (at least earthly) authors, as shown by the following observations: each of the stories uses a different name and description for God, in one case the majestic deity Elohim, and in the other the anthropomorphic being Yahweh; before creation, the world in one story is described as a watery chaos, and in the other as a desert wasteland; in one story man and woman are created simultaneously, and in the other man is created first.

Modern biblical scholars believe that both creation stories were decanted from even more ancient Mesopotamian traditions, which were encoded in written forms as the so-called J-document and P-document. The J-document, the name of which comes from JHWH (a vowel-less term loosely translated as "Lord," though we do not know the precise meaning of the original Hebrew word), is the older, probably dating from the time of David and Solomon (in the tenth century B.C.). This document was composed when Assyria was the dominant kingdom in the cradle of civilization, and it is the source of the folksy creation story we know as Adam and Eve in the Garden of Eden. The younger P-document is thought to have been composed when the land of Judah was destroyed and the Jews were enslaved by the Babylonians (in the sixth century B.C.). The P-document creation story ("In the beginning . . .") is more poetic and liturgical, and thus thought to have been written by priests (whence the "P").

Tradition maintains that Moses wrote Genesis, but there is no documentation or even suggestive evidence for that idea. Instead, many biblical scholars believe that the J- and P-documents were combined into the book we now know as Genesis by a small group of devout Hebrew theologians. These people might be better described as editors,

The carved figures on this Babylonian boundary stone, from the twelfth century B.C., are illustrated for clarity in the drawing on the right. These ancient peoples viewed the universe as a mountain rising from the sea, and the hierarchy from bottom to top on this stone indicated a gradation in divinity. The cosmic gods and goddess at the top (represented by planets and the Moon) were beyond even the reach of prayer, but the lesser gods below them (including Marduk, creator of the universe) were more accessible. Humanity appears just above the bottom rung, and at the bottom of the mountain are the ocean and symbols for the four elemental constituents of the world: earth, air, water, and fire. A snake winds up the side of the stone marker and cannot be seen in frontal view; it is thought to symbolize the chaotic waters that surround the universe, supporting it from below and raining down on it from above. (British Museum)

because they apparently included what they considered to be the most sensible and edifying parts of previous creation stories while omitting other, more objectionable parts. They were also consummate wordsmiths. The measure of their success is that Genesis still moves us, even thousands of years later. But we must not lose track of the fact that Genesis, however inspired, was very much a product of its times—a period preceding the discoveries of science, when Babylonian myths and legends undergirded explanations of nature.

If we carefully scrutinize the P-document, the more scholarly of the two Genesis creation accounts, it is easy to see how the beliefs of the day influenced its final form (if, indeed, the Revised Standard Version of the Bible can be considered the final form for a book so often translated and otherwise modified). In particular, the P-document attempted to define a clear separation between the religious views of the Jews and their Babylonian conquerors, although its authors were not completely successful. The Babylonians believed that the universe was the result of a titanic confrontation between deities: the goddess Tiamat and the god Marduk, who slew Tiamat and from her remains created the world and its surroundings. In the comments following each Scripture passage below, I will point out some of these similarities and omit what modern science has to say about creation, except for noting some scientific concepts that are so obvious as to require no explanation. Since the writers of Genesis chose to discuss the creation in daily increments, I will follow their lead.

> In the beginning God created the heavens and the earth. The earth was without form and void, and darkness was upon the face of the deep; and the Spirit of God was moving over the face of the waters. And God said, "Let there be light"; and there was light; and God separated the light from the darkness. God called the light Day, and the darkness he called Night. And there was evening and there was morning, one day.

The name *Genesis* was applied to the first book of the Old Testament by Greek translators; its proper Hebrew name was *Bereshit,* meaning "In the beginning." Thus the name of the P-document account restates its opening words in the same way that the Babylonian creation epic, the *Enuma Elish,* is named after its first words.

The opening statement seems to imply that the Earth's raw material was present from the outset, though without form, and God's role was to sculpt this matter into the world we know. This notion parallels the view of the ancient Greeks, whose mythology describes initial *chaos,* upon which order (*cosmos*) was imposed. The *deep,* a term describing the vast, restless ocean, provides another way of visualizing the disorder in the beginning. In the Babylonian myth, chaos was represented by the goddess Tiamat, who was thought to be as powerful and as uncontrollable as the raging sea.

Light is the quintessential symbol of order. God created light, whereas darkness already existed. A myth of the Sumerians, the ancient Mesopotamian civilization that preceded the Assyrians and Babylonians, introduced the concept that light and darkness represented the equally strong forces of good and evil, locked in constant battle. In Genesis, God controls and confines, but does not do away with, the darkness. *Day* and *night* are expressly defined (though the Sun does not appear in this account until three days later).

A time scale for creation is also introduced and, lest we misinterpret the length of a day of creation, the phrase *evening and morning* is apparently intended to make it clear that we are dealing with a twenty-four-hour day. In the modern world we begin a twenty-four-hour day at midnight, but in the Babylonian culture, a day began at sunset. Thus, an evening and the following morning were considered parts of the same day.

> And God said, "Let there be a firmament in the midst of the waters, and let it separate the waters from the waters." And God made the firmament and separated the waters which were under the firmament from the waters which were above the firmament. And it was so. And God called the firmament Heaven. And there was evening and there was morning, a second day.

In the Genesis account, the order in which God created objects in the universe is precisely the same as the order in the Babylonian creation myth. First came the *firmament,* which, from a modern perspective, is a very odd term to apply to the sky. The Hebrew word from which it is derived literally means "a thin metal plate," and the writers of the P-document obviously considered the firmament to be a hard, solid dome. The Babylonians pictured the universe as a vast room, with the sky as its ceiling and the Earth as its floor. The solid Earth was surrounded by water on all sides, and on the other side of this moat stood mountains that supported the firmament. The idea of a solid, touchable sky was a common theme in ancient times; for example, the giant Atlas of Greek mythology supported the firmament on his considerable shoulders.

The oceans are separated by this solid, waterproof sheet, so that there is water both above and below the firmament. This notion may have been an outgrowth of life in arid regions, where the Hebrews, Babylonians, and their predecessors farmed. Waters under the firmament, in the form of rivers or the Mediterranean Sea, were familiar enough, but water above the firmament was not. Rain was scarce and, on the rare occasions when it did fall, was probably viewed with some wonderment. Water leaking from above the firmament could logically have been viewed as the source of rain, and might also have provided an explanation for the sky's blue color.

The P-document emphasizes that the firmament did not just magically appear at God's bidding (as did light); instead, after announcing it, God *made,* or fashioned, this material object. The idea of a deity somehow hammering a giant metallic lid and fastening it over the Earth may seem ridiculous or even irreverent, but in the Babylonian creation myth Marduk constructed the world from Tiamat's remains by labor, not by pronouncement.

> And God said, "Let the waters under the heavens be gathered together into one place, and let the dry land appear." And it was so. God called the dry land Earth, and the waters that were gathered together he called Seas . . . and God said, "Let the earth put forth vegetation, plants yielding seed, and fruit trees bearing fruit in which is their seed, each according to its kind, upon the Earth." And it was so . . . and there was evening and there was morning, a third day.

Having created light and the sky, God now turns to the Earth itself, and his first order of business is to make dry land. Presumably, the still-chaotic Earth at this point was comprised of water, perhaps a muddy slurry, which is caused to separate into oceans and land masses that rise above sea level.

Almost immediately, plants appear on the land. There is no explicit recognition that plants also grow in the sea as well, and the order of appearance of land plants and sea-dwelling organisms differs from that in the fossil record. The reason that plants appeared before animals in Genesis is that God saved the creation of "living things" for another day. In the time of the P-document, plants were not considered to be alive in the same sense that animals were. Life required mobility, something rooted plants did not have. Instead, the inanimate kingdom of vegetables was viewed simply as a green veneer of the Earth itself, its chief purpose to make edible seeds and fruit.

> And God said, "Let there be lights in the firmament of the heavens to separate the day from the night; and let them be for signs and for seasons and for days and years, and let them be lights

> in the firmament of the heavens to give light upon the earth." And it was so. And God made the two great lights, the greater light to rule the day, and the lesser light to rule the night; he made the stars also. And God set them in the firmament of the heavens to give light upon the earth. . . . And there was evening and there was morning, a fourth day.

Not until the fourth day did God make the Sun (though day and night existed from the first day), the Moon, the planets, and the stars. Thus, the Earth is older than the rest of the solar system, but only by a few days. The starry and planetary *lights* were set squarely into the firmament. Light already existed, but now it is somehow sequestered into these heavenly objects.

An important function of these lights was to serve as markers for fixed times, such as days, months, or years. The monthly or yearly periods are, of course, distinguished by the motions of the heavenly lights, although Genesis gives no indication of how objects embedded into the solid firmament could move.

The Babylonians had carefully worked out the motions of the planets against the fixed backdrop of stars. Motion is implicit in the word *planet,* which is derived from a Greek word meaning "wanderer." In a society that looked to the heavens for guidance, these rapidly moving lights obviously had to have special status, and so each planet was identified with a specific god or goddess. Babylon was basically a civilization of astrologers who viewed the planetary movements and the shapes of starry constellations as cryptic codes provided by the gods to mankind. Deciphering these codes was a major part of their religious beliefs. The Hebrews scorned their captors' religion and, while adopting parts of the Babylonian creation myth into their own creation story, removed the offensive notion of multiple gods in favor of one transcendent God. If the planets were not deities, then their motions (and the messages they supposedly carried) were unimportant and need not be mentioned in this creation story.

The stars, especially, seem almost an afterthought. Because they gave relatively little light and their motions are not important for establishing a calendar, they were dismissed with a few brief words.

> And God said, "Let the waters bring forth swarms of living creatures, and let the birds fly above the earth across the firmament of the heavens." So God created the great sea monsters and every

This German carving shows God attaching the stars onto the firmament, a sphere enclosing the solar system. The planets are fastened to the inner crystalline spheres, according to the Ptolemaic system. (Verlag Karl Alber, Freiburg im Breisgau)

> living creature that moves, with which the waters swarm, according to their kinds, and every winged bird according to its kind. And God saw that it was good. And God blessed them, saying, "Be fruitful and multiply and fill the waters in the seas, and let birds multiply on the earth." And there was evening and there was morning, a fifth day.

The nonliving part of creation was now finished, and it was time to open the curtain on the most important act—life itself. This, too, was done in stages, leading progressively to the inevitable climax, mankind.

Since humans, obviously, were to live on the land, populating the continents was saved for later. First, God brought forth living things to inhabit the water and the air. The phrase *every living creature that moves* distinguishes these living forms from the rooted plants created several days earlier.

Some scholars have suggested that the *sea monsters* (the whales of the King James Version) may relate to the sea monster of chaos in the Babylonian creation myth. If so, the writers of the P-document are again taking aim at their captors' gods and exerting the superiority of their own. Specifically, they state that even these most monstrous of living things cannot be godlike beings, as depicted in Babylonian mythology, but instead are creations of God and thus subject to his will.

> . . . And God made the beasts of the earth according to their kinds and the cattle according to their kinds, and everything that creeps upon the ground according to its kind. . . . Then God said, "Let us make man in our image, after our likeness; and let them have dominion over the fish of the sea, and over the birds of the air, and over the cattle, and over all the earth, and over every creeping thing that creeps upon the earth." . . . So God created man in his own image, in the image of God he created him; male and female . . . And there was evening and there was morning, a sixth day.

Finally, it is time to populate the land, first with animals. By *beasts,* the writers of the P-document meant wild animals. In contrast, *cattle* refers to any kind of domesticated animal. A literal reading of this section would imply that domesticated animals were created in that state from the beginning; however, there is ample evidence that humans domesticated dogs and cats. *Everything that creeps* encompasses reptiles, amphibians, and insects.

That task finished, God is now ready to undertake the penultimate act of creation, resulting in something clearly more than an animal in the view of the biblical writers. The curious phrase *let us* seems inappropriate in a document written by staunch Hebrew monotheists. There have been many explanations suggested for this plural term (God is talking to his angels; this is a use of the royal "we" to indicate his majesty; God is a trinity of Father, Son, and Holy Ghost), but none of these possibilities was part of Hebrew tradition or thinking when the ancient P-document was composed. A more plausible explanation is that the plural term was inherited from the vocabularies of the Babylonians and more ancient peoples. The Hebrew word *Elohim,* meaning "God," is actually a plural form, but after generations of Israelites had been immersed in or enslaved by polytheistic cultures, "the gods" became such a familiar term that they often used it to refer to their own singular deity. (An interesting aside is that the Hebrew word for "man" is *adam,* which explains the name of the central character in the J-document creation story that follows in Genesis.)

It is unclear how literally the P-document writers intended their readers to understand the concept that man was created *in the image of God.* This could, of course, refer to our singular ability to reason, to know the difference between good and evil, but it may also have referred to actual physical form. In Babylonian and other early religious traditions, gods were often pictured as having human forms. Statues of Greek gods are human in appearance, and it would be easy to be envious of their beauty.

> Thus the heavens and the earth were finished, and all the host of them. And on the seventh day God finished his work which he had done, and he rested. . . .

God's creation of everything took six days, which he followed with a well-deserved day of rest. The choice of six working days also stems from Babylonian culture. With the naked eye, the Babylonians could see six visible wanderers (Mercury, Venus, Mars, Jupiter, Saturn, and the Moon) in the night sky. Each day was governed by a particular god and, over time, an extra day was regularly set aside for religious celebrations. The captive Hebrews were obviously familiar with the seven-day week, and were happy enough to take advantage of the weekly day of rest from their labors. However, since they could not acknowledge the Babylonians' religious justification for six days of work and a day of rest, they developed their own. The writers of the P-document provided the rationale in their creation story.

GREEKS BEARING GIFTS

The Book of Genesis was to define Hebrew thought about creation over the next two millennia, but at the time that the P-document was set down, the center of intellectual activity was already beginning to shift, this time to Greece. It is unclear why Greece blossomed so triumphantly. Some historians have suggested it happened because the Greeks liberated knowledge from the priesthood and transferred it to the laity. The organized knowledge of the Babylonians and of the Hebrews was a tradition handed down through generations of priests, but in Greece new ideas sprang from individuals who did not claim to speak for deities. The Greeks had an insatiable curiosity about the Earth and its cosmic surroundings, which drove them to create the true beginnings of science. At that time astronomers were not so keen to understand the origin of the universe as to catalog it, so philosophers dominated discussions about the origin of the universe.

Anaxagoras (ca. 500–ca. 428 B.C.) was perhaps the most extraordinary of these early philosopher-scientists, though also the most maligned. A rich man, he eschewed his possessions so as to devote himself to the investigation and understanding of the heavens. But his efforts were not driven by piety. He refused to see anything divine in the night sky, instead maintaining that the planets and the Moon were of the same general nature as the Earth and had become incandescent through rotation. Even the mighty Sun he viewed merely as a huge mass of glowing metal. Anaxagoras argued that the universe had begun as a chaotic collection of stones, water, clouds, and air. Within this, a vortex was generated, which spread in ever-widening circles. Each of these components separated out in turn as a result of the circular motion, the heaviest remaining in the center to form the Earth. Finally, in the violence of the ensuing swirls, stones were torn away from the Earth and were kindled into stars. This cosmogeny had much in common with another hypothesis proposed thousands of years later, but it was not much appreciated by Anaxagoras' contemporaries.

In those days the Greeks worshiped a panoply of heavenly deities, mostly helpful beings to whom they looked for comfort and assistance, beings who were susceptible to their entreaties and even their bribes. Anaxagoras tried to replace these gods with masses of rock and metal, and the people of Athens wanted no part of it. So they prosecuted him for impiety and atheism. His fate is unclear: one account suggests that he was convicted and banished from Athens, and another says that he was acquitted but found it prudent to absent himself voluntarily. Whichever is correct, Anaxagoras' persecution was the opening salvo in a long and recurring war between religion and science.

A century after Anaxagoras' passing, Greek culture had begun to decline and, with it, Greek science. The decline was accelerated when Alexander the Great conquered the region, along with most of the rest of the known world. Alexander decided to celebrate his victories by building a new capital at the mouth of the Nile, humbly naming the unborn city Alexandria. These grandiose plans collapsed with his death in 323 B.C., and his kingdom was divided among those who could scavenge and hold pieces of it. Egypt became the property of Ptolemy, one of Alexander's generals, who adopted Alexander's ambitious plan to finish the city as his own. Adjoining his palace, Ptolemy constructed a Temple of the Muses, roughly equivalent to a modern university, and staffed it with the most eminent Greek scholars of the day. This was to become the house of science for almost 1,000 years.

Alexandrian scholarship in mathematics, medicine, and astronomy was remarkable, especially because of a change in method, from sometimes dreamy speculation based on general principles to precise scientific attacks based on observation and experiment. The two scholars of this school who most influenced the understanding of creation were Aristarchus and Claudius Ptolemy (no relation to the earlier builder of Alexandria). The former gave us the first true description of the arrangement of the heavens, which was largely ignored at the time. The latter provided an entirely erroneous description that held sway, almost unchallenged, for nearly 2,000 years.

Aristarchus' profound insight derived from his determination that the Sun was many times larger than the Earth. At that time it was universally accepted that the Sun revolved around a stationary Earth. Aristarchus must have pondered the irony that such a huge object would revolve around another so much smaller, and this apparently led him to make the radical suggestion that the Earth might not be the center of the universe—it and the other planets must move about the Sun. Alas, Aristarchus' giant bound in understanding was simply too far ahead of its time (third century B.C.). It was too hard for his contemporaries to accept that anything as large as the Earth could either move or constitute a lesser part of the universe.

Other, less incendiary Greek philosopher-scientists devised a more acceptable understanding that has come to be called the Ptolemaic system, after Claudius Ptolemy. One of the last of the Alexandrian scholars, Ptolemy lived in the second century A.D. Ptolemy's practical intent was primarily to tabulate the heavens as a navigational tool for ships, but in summarizing Greek astronomical thought, he became the standard bearer for this viewpoint. The Ptolemaic system remained the definitive scientific (that is, nonbiblical) model for creation, and it harassed astronomy until the end of the Middle Ages.

What was this flawed model that so well stood the test of time? Like his predecessors, Ptolemy made a great distinction between the obviously imperfect Earth and the heavens, which he held to be perfect and unchanging. The Sun, Moon, planets, and stars were thought to be fastened to a series of crystalline, concentric spheres that constantly rotated, with the motionless Earth occupying the focus. Although not correct, this was an obvious improvement over the Babylonian view of the universe, which could not explain where the Sun spent the night. Planets other than the Earth, fastened as they were to perfect spheres, were thought to move in perfect circles. Greek astronomers had already noticed that the planets did not move through the sky at constant speeds, and they sometimes edged forward and sometimes backward when viewed against the fixed backdrop of stars. Ptolemy argued that their circular orbits (that is, their crystalline spheres) were not quite centered on the Earth, so that as seen by us the planets seemed to move fastest when they were nearest.

Ptolemy's views also encompassed the materials from which nature was fashioned. He thought that the Earth was composed of four rather mundane constituents—earth, water, wind, and fire—whereas the heavens were made from a perfect fifth element, quintessence. The Earth originated at the same time as the rest of the observable universe but, by comparison with shining quintessence, was too flawed to really be considered part of the cosmos, even though it was situated at its very center.

By the time of Ptolemy, the Roman Empire had greedily established its dominance over the Mediterranean region, and Alexandria was administered by Rome. Greek thought, once so adventuresome, slowly became subservient to its new masters. The Romans were extraordinary moralists, engineers, soldiers, and legislators, but decidedly not philosophers and scientists, and so the knowledge and methods of inquiry acquired by the Greeks languished. But the real danger to science came later and from another quarter.

ASTRONOMY IN THE DARK AND THE DAWN

At the beginning of the fourth century A.D., only a small fraction, perhaps a fifth, of the known world's population was Christian. In the year 312, Constantine the Great, Emperor of Rome, embraced the Christian religion. (This landmark event in human history was not the result of Constantine's moral conversion, but rather because he believed that his use of a Christian emblem had brought him victory on the battleground.) Constantine was an illegitimate son of a Roman officer and a Serbian innkeeper, and his elevation to Emperor was by election of an army in the field. It did not take long before pagan religions were forbidden. Christianity reigned supreme, and virtually all thought was dominated by the priesthood. When the Roman Empire finally crumbled, science became irrelevant, as medieval Europe focused not on understanding the heavens but on earthly salvation and religious mysteries. This was not Christianity's finest hour. A torpor settled upon the human mind, as intolerance and credulity combined to stamp out innovative thought in the Western world.

During this dark millennium, the Arabs assumed the role of curators of the planet's scientific knowledge. Spurred by the rise of Islam, the Arabs started a campaign of military conquest and, with breakneck speed, built a formidable empire. Everywhere they conquered, the Arabs absorbed knowledge as well as territory. They excelled as translators of Greek and Latin texts and, in hindsight, they provided a great service by ensuring that the Hellenic legacy of scientific inquiry was not irretrievably lost.

In the thirteenth century church scholars in the West, especially Thomas Aquinas, began to rediscover Ptolemy's ideas and to forge a link between them and the doctrines of Christian faith. Aquinas argued that the Greeks, though heathens, had revealed God's creation, and he was so convincing that the Ptolemaic system became orthodoxy. Aquinas was a classical scholar, but he was no scientist. He and his successors neither observed nature nor conducted experiments, as the Greeks had done.

The Ptolemaic system appeared to work fine for a considerable time, but after many hundreds of years some planets were noticeably far away from where Ptolemy's tables predicted them to be. This situation can be envisioned as something like a clock that runs a few seconds slow. The clock keeps adequate time for a while, but eventually the daily errors accumulate. A number of thinkers, among them a Polish churchman named Nicolaus Copernicus, were beginning to worry that the "official astronomy" might be flawed. Copernicus noticed that the planetary motions could better be accommodated in a model by which the Earth moved too, and he labored for many years to refine this model so as to agree with all the available observations. His model necessitated the relocation of the Sun at the hub of the planetary system, a conclusion already reached 1,000 years earlier by Aristarchus. Locked into the notion that circular planet motions were inevitable, Copernicus replaced Ptolemy's circular orbits with a new set of circles whose centers did not coincide with the Sun. Of course his model passed all the observational tests, for he had introduced one complication after another to assure that it would. The Copernican system was certainly a step foward, for in this model the Sun occupied its rightful place at the center, but the circular planetary orbits were little improved from Ptolemy's.

Once the work was done, Copernicus shrank from the task of preparing it for publication. Poland was also far removed from any printing center that could profitably handle a book so technically complex, so Copernicus allowed his enthusiastic disciple Georg Rheticus to carry a copy of the manuscript (*De revolutionibus orbium coelestium*) to Germany for printing. Unfortunately, Rheticus left it in the unsupervised care of the printer and his clergyman-proofreader, who conspired to insert a new introduction: "These hypotheses need not be true nor even probable. . . . Let no one expect truth from astronomy, lest he leave a bigger fool than when he entered." Copernicus only saw the finished book in 1543 as he lay on his death bed, where he is reported to have been agitated by this unauthorized addition to his work. The collapse of the Ptolemaic system did not follow its publication very quickly, as Copernicus' hypothesis required that church leaders wrestle with a new concept. After all, its central conclusion, that the Earth was not the center of the universe, led to an unsettling corollary: If mankind was the climax of all creation, as clearly implied in Genesis, then he had been assigned a home in space that was not commensurate with his exalted position. Not surprisingly, Copernicus' great work soon made the Roman Catholic *Index of Forbidden Books*.

Just a few decades later, an eccentric but talented Danish astronomer named Tycho Brahe began to compile observations of the motions of planets in the night sky. The precision of his measurements was amazing, considering that he made them with the naked eye. Brahe opposed the ideas of Copernicus, believing them contrary to sound science and scripture. Accordingly, he set himself the task of improving upon and shoring

The Earth-centered Ptolemaic solar system, on the right, hangs more heavily on the balance of truth than does the Sun-centered Copernican system, on the left, in this frontispiece from a book written a century after Copernicus published his work. (From J. B. Riccioli's *Almagestum Novum*, published in 1651)

up the Ptolemaic system and, needing assistance, he hired Johannes Kepler. Upon Brahe's death in 1601, Kepler inherited his observational data, which formed the basis for his revolutionary calculations of planetary orbits. Kepler ignored astronomy's obsession with the circle as nature's own curve, and found that planetary observations better fit elliptical orbits. Kepler's pioneering work firmly buttressed the idea of a Copernican system and paved the way for Isaac Newton's brilliant explanation of gravity as the engine of celestial mechanics.

The triumph of these discoveries also led to the popular notion that the solar system could be viewed as a clockwork, a perfect mechanism on a gargantuan scale. This view sent mechanically minded craftsmen into a flurry of activity constructing ornate, gear-driven models for the solar system, called "orreries." Such beautiful instruments became extremely popular, and the perception of a clockspring solar system naturally led to the belief that the construction of such a well-ordered system of orbiting bodies surely required the hand of a divine clockmaker.

As far-reaching as Newton's work was, it did not really address the problem of how the solar system originated. Others, however, did seek answers to this question. One popular idea was put forward by Georges-Louis Buffon in 1778. Buffon proposed that a catastrophic collision between the Sun and a comet ejected a thread of matter that condensed to form the planets. When comets were later shown to be small objects, the idea languished until the colliding comet was replaced with a passing star.

Newton had commented on the fact that all the planets move in the same direction around the Sun and that their moons all move in this same direction around the planets, a regularity he attributed to the intent of the Creator. In 1796, Pierre Laplace, who has sometimes been described as the French Newton, published his contrary view that this regularity was due to natural causes. In his "nebular hypothesis," Laplace argued that an initial nebulous mass of hot gas had flattened into a disk as it rotated, and shrunk as it cooled. When it could flatten no further, it broke into pieces by shedding ring after ring of matter, which finally condensed to form the planets.

A CONFRONTATIONAL TEA PARTY

As the nineteenth century opened, geology and biology were just beginning to emerge as formidable sciences, destined to put new wrinkles in the old arguments about creation. The fact that the Earth, the abode for humankind, was a relatively small lump in a very large cosmos was reasonably well established, and religion had made its peace with that revelation. However, new arguments about mankind's place in the immensity of geologic time, as well as our biologic relationship to the rest of the living world, were soon to mushroom into open warfare.

At the beginning of the last century, most scientific interpretations were still made in the biblical context. Scientists and nonscientists alike accepted the notion that the elegant complexity of the Earth and especially of its life demanded acknowledgment that it had been *designed* (a reiteration of the earlier, astronomical theme that a watch requires a watchmaker). Likewise, the biblical Flood that floated Noah's ark was no fable but a real deluge that had to be accommodated in explanations of geological strata and the fossils they contained. These were not beliefs forced upon a fledgling science by the church but, for the most part, willingly accepted constraints fortified by the best theological and philosophical arguments of the day. However, nineteenth-century England was destined to become a cradle for innovative naturalists whose thoughts were to disrupt this staid and comfortable worldview.

Some of the most important figures of these debates—geologist Charles Lyell, biologist Charles Darwin, physicist Lord Kelvin—were British contemporaries; they could have had tea together and chatted about Genesis and science. We can only speculate as to the content and tenor of such an afternoon's conversation, but a review of the ideas and considerable contributions of each of these scientists may provide us with some insights.

Geology's guiding principle at this time was *catastrophism,* the idea that the Earth's history had been punctuated by sudden events. These catastrophes, although originally envisioned as geological, were rather quickly reshaped into biblical events such as the Flood. Most geologists were convinced that Genesis and geology could be made compatible, and they worked hard to make it so. Charles Lyell's own teacher at Oxford, William Buckland, was particularly influential in convincing his contemporaries that the object of geological research was to gather evidence that supported Scripture or proved the existence of God. Finally forced to concede that the formation of geologic strata demanded a complex history considerably longer than six days, Buckland argued that the rock record must have been established *before* the events depicted in Genesis. In other words, all of geologic history was encompassed within the biblical phrase "In the beginning," and the study of rocks allowed him to see what the Earth might have been like before God's intervention.

Charles Lyell differed from his mentor in arguing that the Earth's history was dominated by the same kinds of geological processes that we observe today, operating at basically the same slow rates as they do in the modern world, a view now termed *uniformitarianism.* But Lyell, too, interpreted geologic observations in a theological context. He explained away (at least to his own satisfaction) the variations in fossils found in strata of different ages by asserting that the Earth and its life forms had not changed progressively but simply oscillated around some mean value with no net change. He felt that the fossil record was imperfect (which it is), and thus gave only the appearance of change (which it does not). Moreover, he was struck by the localized geographical distribution of fossil plants and animals, and argued that they must have been specifically created by God in that locality. The Noachian Flood, however, was not part of Lyell's geology, perhaps because it would not qualify as a uniformitarian event.

Few of Lyell's contemporaries accepted his peculiar view that life was unchanging, and even he abandoned the idea eventually. But no one had a plausible explanation for just how individual species came to be, save for the scriptural demand that God had created each and every one. In the 1830s this question was of sufficient import to be called "the mystery of mysteries."

But this mystery did not remain vexing for long. The answer came as an unexpected outgrowth of a round-the-world cruise by the H.M.S. *Beagle,* which carried as its passenger one Charles Darwin. In 1831, Darwin sailed away from England believing in catastrophism and unchanging species. But all that was to change dramatically. Darwin carried with him on his voyage a copy of Lyell's *Principles of Geology,* and in reading it he became a convert to uniformitarianism. This view of a slowly but progressively changing physical world undoubtedly influenced his perception of the living world.

As the *Beagle* skirted the coast of South America, Darwin recorded and described species gradually giving way to closely allied species. In Argentina, he unearthed the fossil remains of an extinct animal similar to the modern armadillo, and this discovery caused him to ponder the relationship between past and present life. He concluded that gradual changes in organisms could occur not only geographically, but temporally. Such changes could be the natural result of changing environmental conditions. These ideas were reinforced by his observations in the Galapagos, where each isolated volcanic island housed its own unique assortment of species that differed only slightly from those of the island next door. Darwin thus came to the inescapable conclusion that life evolved.

After returning home, Darwin ruminated for years on what might cause evolution before finally stumbling across an answer: natural selection. Mother Nature clearly made more organisms than she could support, with the brutal consequence that many organisms died before they could reproduce. Those organisms that lived on were not just lucky, but happened to be endowed with traits that enabled them to compete more effectively in an unforgiving environment. They tended to pass on these traits to their offspring, and in so doing forced gradual changes in the species population. Eventually these accumulated changes were great enough that it could be said that one species had evolved into another.

Humans, in Darwin's view, must also be the result of evolution—just the tip of one random branch of a complex evolutionary tree, and perhaps not the culmination of God-guided, unidirectional change. Obviously, this put Darwin at odds with the views of most of his fellow countrymen. In fact, it eventually undermined his own religious beliefs, which evolved from theism to agnosticism during the decades after the publication of his ideas.

Evolution by natural selection also put Darwin in conflict with the foremost British scientist of his day, Sir William Thomson, later known as Lord Kelvin. Much of Kelvin's research dealt directly or indirectly with heat and, in fact, he later lent his name to a unit of temperature widely used in science (the temperature interval corresponding to one degree in the Celsius scale is called a "kelvin" in the absolute temperature scale). In 1865, Kelvin dropped a two-page bombshell brazenly titled "The Doctrine of Uniformity in Geology Briefly Refuted." This publication attacked Lyell's uniformitarianism by arguing that the Earth has continually cooled from an initially hot state since its formation, so that it could not have existed for a very long time under uniform conditions. In particular, Kelvin felt that the Earth was not old enough to have allowed natural selection to force the evolution of its life forms. Instead, he saw God's design, rather than random evolution, in human consciousness. Kelvin, like many of his predecessors, was enamored of the concept of God as clockmaker for the rest of creation, which led him to comment: "There may in reality be nothing more of mystery or of difficulty in the automatic progress of the solar system from cold matter diffused through space, to its present manifest order and beauty, lighted and warmed by its brilliant Sun, than there is to the winding up of a clock, and letting it go till it stops."

To gain an appreciation for the influence these scientists wielded in their day, one only needs to visit Westminster Abbey, the very embodiment of British history and culture. On my first visit to this magnificent cathedral, I remember standing awestruck near the entrance, trying to take it all in. Several minutes elapsed before I glanced down and realized, to my horror, that I was standing on Lyell's crypt. To be sure, it is difficult to walk anywhere in the Abbey and not tread on the dead, but it seemed particularly odious for a geologist to be standing on the grave of the father of uniformitarianism. Continuing a few steps farther down the nave, just across from kings and prime ministers, I soon found the remains of Darwin and Lord Kelvin, ironically almost side by side. I doubt that many visitors to Westminster Abbey recognize these scientists' names (save Darwin), but this shrine is reserved only for those who shaped history, and these gentlemen qualify.

A tea party attended by Lyell, Darwin, and Kelvin might have been interesting indeed. Genesis would certainly have had its champions from among these revered scientists, and "one lump or two" may have described more than how they took their tea. We can only guess as to whose arguments might have carried the day.

THE TEA PARTY CONTINUES

Surprisingly, some of the same arguments that Lyell, Darwin, and Kelvin would have proffered at this tea party continue to be made more than a century later, especially in the United States. Sporadic skirmishes break out between Christian fundamentalists ("creationists") and scientists, sometimes with high stakes.

The foundation of the fundamentalist movement was the publication in the early part of the twentieth century of a set of twelve booklets that argued for divine design and biblical inerrancy. Shortly thereafter, the theological dispute spilled over into the public sector, when serious efforts began to block public schools and universities from teaching evolution and other scientific theories that were thought to be incompatible with traditional interpretation of the Bible. The single event that most symbolized this struggle was the celebrated trial, in 1925, of *Scopes v. Tennessee,* resulting from the indictment of a high school teacher for discussing evolution in the classroom. The Tennessee legislature had just enacted a bill prohibiting the teaching of "any theory that denies the story of Divine Creation of man as taught in the Bible and to teach instead that man has descended from a lower order of animals." The case itself was not unique, because Florida and Oklahoma had passed similar legislation, and other states were soon to follow. It did provide notoriety, however, because of a confrontation between two especially able protagonists, William Jennings Bryan and Clarence Darrow. Although prosecutor Bryan prevailed and Scopes was convicted, the biblically grounded attack on evolution was largely discredited in the attending press coverage. Nevertheless, the "monkey law" was upheld by the Tennessee Supreme Court and remained on the books until 1968, when the United States Supreme Court struck down an Arkansas adaptation of the Tennessee statute on the constitutional grounds of separation of church and state.

As a legal issue, this is a debate that will just not go away. In 1982, a federal district court in Arkansas struck down a recently enacted state law mandating a balanced treatment of creationism and evolution in the public schools. During the hearing, the statute was attacked by a dazzling array of church and educational groups, and was supported by an equally vociferous group of creationists. In the end, this law, too, was found to violate the establishment of religion clause of the Constitution.

As I write this, several current events have reinvigorated public interest in creationism versus science. In 1996 the Tennessee state legislature narrowly averted passage of a bill that would allow any teacher presenting evolution as fact to be fired for cause. But my own state is not unique in its attacks on science. During the same year, the National Broadcasting Company aired a controversial prime-time television program suggesting that the scientific establishment was suppressing evidence that humans coexisted with dinosaurs, a favorite theme of creationists.

Whenever religion and science square off to do battle, it is not a pretty sight. In truth, both sides are often at fault. Dogma usually wins, regardless of whether religion or science prevails. Religious fundamentalists tend to appropriate God to defend their special view of creation and to label any who disagree as infidels. In response, scientists often cite arcane research data as supreme authority and not infrequently characterize those who hold contrasting views as ignorant. (In noting that the arguments of both sides can sometimes be needlessly uncivil, I am not conceding that the two views are equally correct.) For some, the gulf between the biblical and scientific views of creation has not been spanned over the centuries.

The creationists' views, as articulated in court cases and in their literature, can basically be summed up succinctly as follows: The Earth is young, no more than a few thousand years old. All living and extinct creatures were created at the same time and must have lived contemporaneously. Catastrophism adequately explains geologic history. The biblical Flood inundated the entire Earth, extinguishing all life forms not saved in Noah's ark. Fossils are the remains of organisms that perished in the flood. Their arrangement in geologic strata has nothing to do with the time in which they lived, because all the strata were deposited essentially in the same instant. Rather, their stratigraphic positions reflect the depth in the ocean at which they lived.

Modern geology, of course, takes great umbrage at each of these views. To see why the beliefs of scientists differ from those of creationists

This cartoon makes the point that six days was not long for creation. (*Frank and Ernest*, reprinted by permission of Newspaper Enterprise Association, Inc.)

requires that we examine how the two groups approach the problem. The differences could not be more dramatic.

Fundamentalists accept, on faith, that the biblical account of creation is literally true. The Bible becomes, in effect, a filter through which scientific data must pass. The acceptability of scientific evidence is judged on whether it supports or challenges a literal reading of the Genesis account.

In contrast, scientists espouse a set of beliefs that are constantly refined (or sometimes even upended), based on rigorous application of the scientific method. The steps of this method—make observations, offer reasonable hypotheses to explain the observations, test the hypotheses, modify or discard them as necessary—can be applied, even to profound questions like the origin of the Earth and its inhabitants. The important distinction to be made here between science and religious beliefs is that scientific hypotheses are falsifiable. Science may not always be right, but it has a built-in mechanism for self-correcting its mistakes. Ideas that cannot pass the muster of continued testing and observation are discarded. In fact, scientists tend to be a rather hard-nosed lot who are loath to embrace any new hypotheses until a convincing body of supporting observations exists, and many seem to delight in overturning others' hypotheses. The question of creation is a difficult one for science, because in many cases the relevant experiments and observations cannot be done. It is not, however, an insurmountable one.

The creation of stars and starfish, of worlds and worms, of matter and mind, often seems wondrous. Both myth and science speak of these same subjects, but the focus of myths is on value and meaning, whereas that of science is on facts. Science does not have all the answers, or even a corner on the market as the only viable route to understanding. For all its mistakes and shortcomings, however, it does offer the most direct access to certifiable truths. Albert Einstein once said, "One thing I have learned in a long life: That all our science, measured against reality, is primitive and childlike—and yet it is the most precious thing we have."

This book attempts to summarize current scientific research and thinking on questions about the origin of the Earth, its cosmic neighborhood, and its inhabitants. By focusing primarily on our own world, we can use insights from a variety of scientific disciplines, such as geology, chemistry, and biology, rather than relying only on astronomy, to explore our origins. If you persist to the end, you will see that our small blue planet is very special, a tiny cauldron of creation whose importance is all out of proportion to its size. The picture of our beginning that emerges is not one with which Ptolemy, Kepler, Lyell, or Kelvin might have been comfortable, but the reality (at least as currently perceived by science) is as miraculous as the myths.

Some Suggested Readings

Asimov, Isaac (1981). *In the Beginning*. Crown, New York. *The first sixty-six pages of this book, by a talented author known primarily for science fiction, provide an interesting and easily digestible discussion of the P-document account in Genesis.*

Atkins, P. W. (1981). *The Creation*. Freeman, Oxford, UK. *The origin of the universe, from a physics perspective; this is an elegantly written little book that focuses on a different set of topics and approaches from the present book.*

Bartusiak, Marcia (1993). *Through a Universe Darkly*. HarperCollins, New York. *Chapter 1 of this engrossing scientific history provides a fascinating summary of the contributions of the ancient Greek astronomers and philosophers, and Chapter 2 describes the Copernican revolution.*

Jeans, James (1948). *The Growth of Physical Science*. Cambridge University Press, New York. *The science in this book is understandably dated, but the writing is exquisite and the author's grasp of the history of science is staggering.*

McGowan, Chris (1984). *In the Beginning: A Scientist Shows Why the Creationists Are Wrong*. Prometheus, Buffalo, NY. *A well-written but somewhat contentious argument against creationists' beliefs about the biblical Flood and evolution.*

Santillana, Giorgio de, and Dechend, Hertha von (1969). *Hamlet's Mill*. Gambit, Boston. *A fascinating interpretation of worldwide myths and their relationships to astronomy.*

Sproul, B. C. (1979). *Primal Myths*. Harper & Row, New York. *If you did not get your fill of creation myths in reading this chapter, this book is for you.*

Wilson, D. B., ed. (1983). *Did the Devil Make Darwin Do It?* Iowa State University Press, Ames, IA. *A volume of essays by scientists and religious scholars that deals with the conflict between creationism and science; the book also ends with contributions summarizing the historical and legal perspectives of creationism in public schools.*

Stardust and Antique Elements

Long Before There Was a Solar System

DUST TO DUST

Like every new cadet at The Citadel, I quickly became fanatical about dust. At this military college, a dusty windowsill or bed frame was punished with demerits, which meant confinement to your room on weekends or, if you had accrued enough demerits, hours of walking briskly back and forth across the barracks quadrangle with a rifle on your shoulder. Every Saturday morning, an upperclassman wearing clean white cotton gloves barged into my room, ostensibly to inspect for military conformity but really, in my mind at least, to seek dust. A second cadet armed with pen and clipboard hovered at his elbow, ready to record any infraction that might result in demerits. Bracing at rigid attention, I prayed that I had removed every offending mote. The inspector would wipe his glove across various flat surfaces and, if he discovered dust, he would hold up his ever so slightly soiled glove inches from my face, and a smile would crease his own. Believe me, I know what it is like to search out and collect tiny dust particles, and I have the utmost respect for people who do it for a living.

Mineralogist Mike Zolensky curates dust for NASA at the Johnson Space Center in Houston. Superficially, Zolensky's miniature collection looks like dust anywhere and, if spread on a Citadel bookcase, would still garner demerits. His dust must have some special significance to deserve the special interest of a space agency. In fact, it is *interplanetary dust,* tiny particles that once orbited in the open spaces between worlds. These miniature grains are collected onto sticky plates mounted on the leading edges of the wings of stratosphere-cruising aircraft. Upon their return to Earth, the individual dust particles are removed and catalogued, after which they are made available to scientists for study. Most of the interplanetary dust collection was originally part of asteroids or comets, in effect free, albeit tiny, samples of other bodies in the solar system. As the particles spiraled inward toward the Sun, some small fraction of them was intercepted by the Earth. It is possible that, hidden among this assortment of interplanetary dust, there may be motes of *interstellar dust,* matter that once floated between the stars. However, only a few microscopic bits of interstellar matter, called GEMS, have been recognized as inclusions within interplanetary grains.

At the University of Chicago, chemist Roy Lewis amasses a dust collection of his own. Lewis stews samples of a particular type of meteorite, called a *chondrite,* in caustic acids. Remarkably, the residue left after dissolution of chondrites consists of interstellar matter. Separating minute interstellar grains in this manner is tedious work, and the end product is a mere thousandth of the mass of the original meteorite. But it's worth the effort. The chondrites themselves are actually aggregates of materials from the earliest solar system, and the fact that they are made in part of interstellar dust allows a glimpse of the way that the Sun and planets were originally fashioned from primal stuff. As it turns out, though, even this interstellar matter is not truly primordial, but was formed many billions of years ago in previous generations of stars. Before we see just how the interstellar dust formed and was dispersed into space, let's go back to the beginning, the very beginning.

START WITH A BANG

One might think that our solar system was unpopular. Everywhere around us, galaxies of stars are rushing away, and their redshifted spectra indicate that those that are farthest away are traveling fastest. Of course, we are not situated at the focus of the universe, as our ancestors thought, no matter how much it may appear so. Our solar system, too, is in motion away from its cosmic neighbors. The expansion of the entire universe suggests that at some time in the past all matter and energy must have been contained together in one place in some ultradense state. Then, perhaps 10 to 20 billion years ago, some kind of gargantuan explosion created more conventional matter and sent it racing in all directions. This idea was originally championed by Belgian astronomer Georges Lemaître, who characterized it in graphic terms:

> The evolution of the world can be compared to a display of fireworks that has just ended: some few red wisps, ashes, and smoke. Standing on a cooled cinder, we see the slow fading of the suns, and we try to recall the vanished brilliance of the origin of the worlds.
>
> —Georges Lemaître, *The Primeval Atom*

Lemaître's explosion, pejoratively called the "Big Bang" by its original detractors, represents the beginning of the universe. As it expanded the universe cooled, and the echoing aftermath of the explosion is a kind of lingering pervasive static, a background radiation that is colder than liquid helium.

We do not actually know how powerful this explosion was, and this remains an important cosmological question. If the Big Bang were truly huge, all matter would have gone flying off into space at very high velocities, only a little slowed by mutual gravity. In this case the expansion might continue forever. Alternatively, if the explosion were somewhat less powerful, the matter now traveling outbound might eventually slow to a halt and then reverse, collapsing back inward. The existence of a universe that expands and contracts again would allow the possibility that time zero for "our" universe might only mark one of several beginnings of time.

The matter emerging from the Big Bang arose from a stew of *neutrons* (nuclear particles with mass but no charge) that has been called *ylem* (pronounced "i-lem"), a term coined from a Greek word for the substance from which all matter was supposedly derived. The matter newly created from ylem was an almost pure mixture of the simplest elements: hydrogen, consisting of both normal hydrogen and its heavy isotope (deuterium); and helium. The two *isotopes* of hydrogen have different numbers of neutrons in their nuclei, and hence different masses. An enthusiastic cheerleader for the Big Bang model whimsically described the creation of the first elements this way:

> In the beginning God created radiation and ylem. And ylem was without shape or number, and the nucleons were rushing madly over the face of the deep. And God said: "Let there be

mass two." And there was mass two. And God saw deuterium, and it was good. . . .

—George Gamow, *The Creation of the Universe*

The gaseous atoms of hydrogen and helium eventually clumped together by mutual gravitational attraction into huge clouds, and ultimately clots of gas were fashioned into the first stars. Nevertheless, this was a pretty dull universe with rather simplistic matter, at least from a chemical perspective. The rich variety of elements now present in our own star and in our world required other mechanisms for formation.

ELEMENT FACTORIES

As a boy, it was my chore to keep the woodpile stocked with logs and kindling. Anyone who has chopped firewood and lugged it to the fireplace soon appreciates that mass can be converted into energy. Einstein's famous equation rigorously describes the equivalence of matter and energy, an insight that explains how the stars shine.

In 1913, Princeton astronomer Henry Russell constructed a plot of the luminosities of stars versus their spectral types (the latter are roughly equivalent to their temperatures). Danish astronomer Einar Hertzsprung had independently done the same thing several years earlier, but it was Russell who popularized this diagram as a way of picturing stellar evolution. The diagram now carries both astronomers' names, although its abbreviated title, the "H-R diagram," is most often used.

Both Russell and Hertzsprung noticed that most stars plot in a band that cuts a swath from the upper left to the lower right of this diagram, a grouping now called the *main sequence*. Stars on the main sequence increase in size progressively from the lower right to the upper left. The stars that lie above the main sequence are *giants,* because for two stars having the same temperature, luminosity increases with size. Conversely, the stars below the main sequence are *dwarfs*.

After trying to understand the significance of this plot for more than a decade, Russell proposed that stars evolve from right to left across the diagram. In his view, the main sequence represents a period of stasis where stars remain for eons during which they burn some unknown nuclear fuel, that is, they transform their own matter into energy. We now recognize that infant stars, called *protostars,* do enter the diagram by doing a complex little dance just to the right of the main sequence before

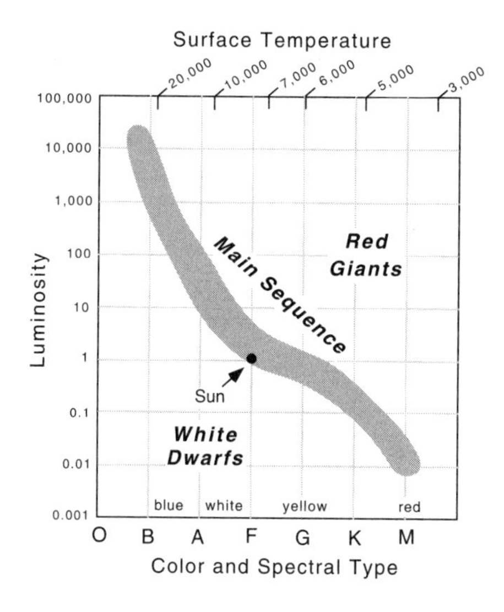

The Hertzsprung-Russell (abbreviated H-R) diagram shows how the luminosities of stars relate to their temperatures and colors. Most stars, including the Sun, plot in a diagonal band called the main sequence, where they remain so long as they burn hydrogen in their cores. After the hydrogen fuel is exhausted, they evolve into red giants, where helium and other elements are burned. Once all the available fuel is gone, they either collapse into white dwarfs or explode.

firmly settling on it as full-fledged stars, but they later evolve from left to right. Although his evolutionary path was faulty, Russell was correct in arguing that nuclear reactions make stars shine. The discovery in the 1930s that stars are mostly hydrogen allowed scientists to infer that the unknown nuclear reaction must involve this element and ultimately to determine how hydrogen is burned in main-sequence stars. In 1952, discovery of the presence in stars of highly radioactive technetium, a rare element whose most stable isotope lasts less than 1 million years, encouraged research on other ways to fabricate elements within stellar furnaces.

We now know that stars are actually nuclear reactors whose formidable luminosities are powered by fusing together many atomic nuclei. Our Sun, like other stars on the main sequence, furiously converts hydrogen into helium in its core. This is pure alchemy, the transformation of one element into another, just as the ancients dreamed. The alchemists were wrong, however, in assuming that they could convert lead to gold under the heat of a candle flame; it takes many millions of degrees, temperatures characteristic of the interiors of stars, to forge new elements from old. At these extreme temperatures, hydrogen atoms are stripped of their electrons, and the naked protons undergo jarring collisions that weld them together.

Our Sun will remain on the main sequence for another 5 billion years or so, until its central store of hydrogen runs out. Once it has exhausted the hydrogen in its core, nuclear burning will then begin to attack a shell of hydrogen surrounding the inert center. The core will then contract and give up gravitation energy, causing the outer envelope of the Sun to expand. At this point our star will dramatically swell into a bloated red giant, jumping horizontally to the right in the H-R diagram. As the burning shell heaps more helium onto the banked core and its temperature increases further, eventually the helium itself will ignite, providing a new fuel source that will fuse into carbon and oxygen. But once the helium is gone, our Sun, a star of only modest size, can evolve no further. It will then shed its outer layers and shrink into a tiny white dwarf, a dense star no bigger than the Earth itself.

Some stars are huge in comparison to the Sun, perhaps twenty times as massive and forty thousand times as luminous. Such stars evolve much more quickly than the Sun, running through their main-sequence existences in 1 million years or less. These leviathans are born, lead full stellar lives, and die while their smaller kin are still forming. At the red giant stage (so called because their outer layers are cool and red), such massive stars aspire to a more complicated evolution of nuclear reactions. Carbon is transformed into oxygen, magnesium, sodium, and neon, while neon, in turn, is transformed into more oxygen and magnesium, and so on, as the ashes from one cycle become the fuel for the next. Because the temperature is higher in the core than at the surface, a huge star eventually becomes layered like an onion, with each layer the locus of a particular nuclear fusion reaction. This progression cannot continue forever, however, because the mutual repulsion of protons (nuclear particles with mass and a positive charge) becomes greater as the nuclei of newly fabricated elements increase in size. Extremely high temperatures, billions of degrees, are needed to make large, heavy nuclei, but the collisions between nuclei are so violent that as many atoms are destroyed as are created. Eventually a kind of equilibrium is achieved, in which only the most stable elements thrive. Iron and its neighbors in the Periodic Table are very stable and are the heaviest elements that stars can fabricate, at least under normal circumstances.

AND FINISH WITH A BANG

It is not possible to forge elements heavier than iron by simply increasing the temperature to ever-higher levels. At about 5 billion degrees, destruction of nuclei wins out over construction of new ones, and so a red giant at this temperature would revert to a ball of helium if it could be held together long enough. But heavier elements—like gold, platinum, and uranium—certainly exist, so there must be some other way to form them.

The solution to the quandary of how to make heavy elements is to pelt atoms with storms of neutrons. Adding too many neutrons to a nuclide makes it unstable, but if the neutrons can be grafted on fast enough (that is, if neutrons are added faster than the resulting unstable isotope can decompose), very heavy isotopes can be generated. Once the blizzard of neutrons subsides, decay of these highly unstable nuclides will produce heavy elements that are otherwise unobtainable. Particle accelerators, operated by nuclear physicists, have managed to transform one element to another by bombarding it with neutrons, and have even created heavy elements so unstable that they do not occur in nature. But where can we go in space, or on the H-R diagram, to find charging hordes of neutrons?

Unlike small stars like our Sun, red giants with greater mass do not just fizzle out into dwarfs when they exhaust their fuel; they are ill-tempered, and forcing a strict diet upon them has catastrophic consequences. Such stars explode with almost incomprehensible violence, an event known as a *supernova*. Within a span of a few minutes temperatures reach extraordinary levels, and the center of the star implodes to become a concentrated mass of neutrons (sometimes called a pulsar). Additional matter falling onto this compressed nucleus rebounds, and when the shock wave reverberates to the surface the star brightens and explodes. For a few short seconds during a supernova event, the interior of the star is raked by a blitzkrieg of neutrons, a place where heavy elements are efficiently manufactured.

Another chore I had when growing up was to remove the ashes from the fireplace and spread them in the flower beds. A supernova explosion likewise cleans out the stellar furnace, redistributing its ashes in the star's backyard. The star's outer layers, containing their booty of newly created heavy elements, are blown with great force into the interstellar medium.

The expanding gas from a supernova explosion cools as it travels

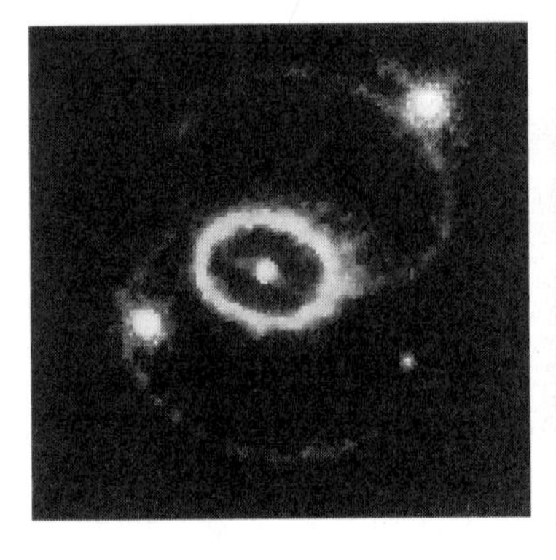
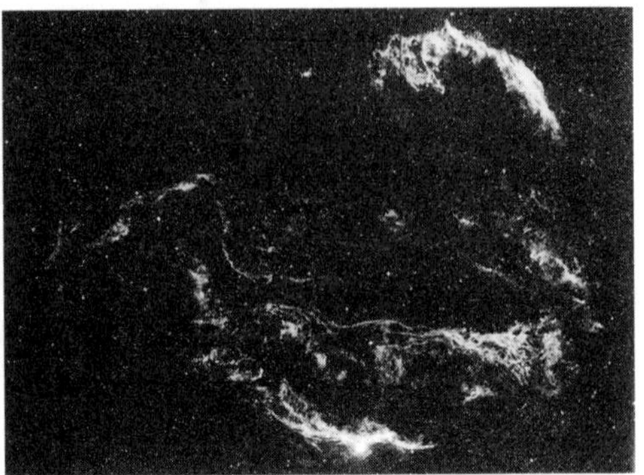

Two views of supernovae: The photograph on the left taken by the Hubble Space Telescope, shows rings of glowing gas and dust expanding outward from a supernova in the Magellanic Cloud that was first noted in 1987. The photograph on the right shows the remnant of a supernova explosion that occurred about 50,000 years ago. This burst bubble, called the Cygnus Loop, is now nearly a trillion kilometers across. (NASA)

outward. Some of the atoms begin to combine into tiny solid particles, essentially interstellar dust. This material eventually finds its way into clouds of hydrogen and helium, which are cocoons for the next generation of stars. That generation will begin its existence seasoned with some amount of heavy elements, a spicy inheritance to which it will, in time, add even more.

COSMIC CHEMISTRY

This amazing concept, of new elements forged inside stars by progressively fusing lighter elements or by bombarding nuclides with hordes of neutrons, was born essentially whole and published in one remarkable scientific paper. In 1957 Margaret Burbidge and her husband Geoffrey, along with William Fowler and Fred Hoyle, teamed together to write an article entitled "Synthesis of the Elements in Stars." In this seminal work, more than a hundred pages long, this team of astronomers and physicists laid out virtually the entire theory of stellar nucleosynthesis. Such groundbreaking work is the modern equivalent of the contributions of Copernicus or Kepler, a paper so influential that it has come to be known to scientists simply as "B^2FH" (from the initials of the authors' surnames), as if it were some universally recognized chemical formula.

The B^2FH model finds elegant confirmation in the *cosmic abundance*

Cosmic Abundance of the Elements

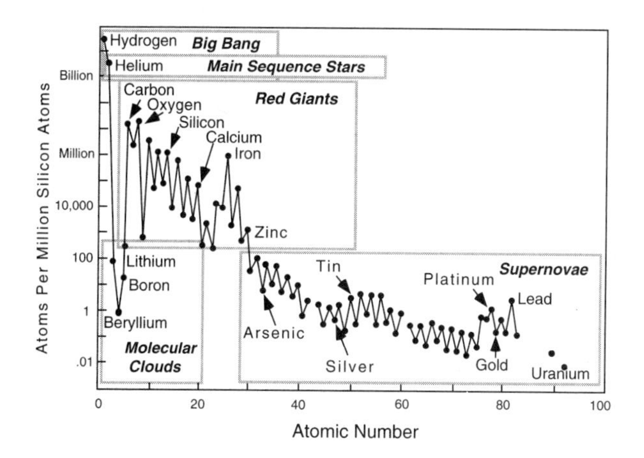

The atomic abundances of the various elements in the universe, each relative to 1 million atoms of silicon, are illustrated here by dots connected with a jagged line. The elements are ordered from left to right according to increasing atomic number (the number of protons they contain, which fixes their identities as elements). Each marked division on the vertical scale signifies a tenfold increase in abundance. The interior boxes enclose elements that formed in a similar astrophysical setting.

of the elements, a tabulation of the chemistry of the universe. Applying the term *cosmic* is somewhat of an overstatement, because in reality we only have reasonably precise measurements of the chemistry of our own solar system, but that composition is enough like that of other stars that it may be a fair approximation of the cosmos. In 1956 chemists Hans Suess and Harold Urey tabulated the natural abundances of elements and isotopes. Their results served as an excellent map for tracing out the nuclear processes proposed by B^2FH, and was undoubtedly crucial in formulating this understanding.

Following Suess and Urey, the variations in cosmic elemental abundances are normally plotted on a diagram versus increasing atomic number (scientists' jargon for the number of protons in the nucleus), which fixes their elemental identities. We now recognize that the very shape of this distribution of elements is dictated by the nuclear processes that formed them. Hydrogen and helium, the products of the Big Bang, constitute about 99 percent of all matter in the universe. More helium is created by burning hydrogen in main-sequence stars. The next elements in line of increasing atomic number—lithium, beryllium, and boron—have very low abundances, because their nuclei are among the least stable in stars. Why they exist at all is somewhat of a mystery and the subject of debate among astrophysicists. It is possible that these elements were formed not in stars, but in the relative emptiness of interstellar space, during collisions of other atoms with cosmic rays.

The rest of the Periodic Table, really the major constituents of planets and of life, constitute only about 1 percent of the cosmos. These elements become increasingly rare as atomic number increases. For example, hydrogen atoms are a million times more abundant than manganese and a trillion times more abundant than gold. The heavy elements, "metals" in the parlance of astronomers, can be lumped into groups, depending on how they formed. The elements from carbon to calcium were fabricated by burning helium and other elements inside red giants. The pronounced abundance peak at iron results from the exceptional stability of this element relative to others as the stars aged. Elements heavier than the iron peak formed by the very rapid capture of neutrons during supernovae.

Finally, the peculiar sawtooth pattern, in which elements with even atomic numbers are much more abundant than the adjacent elements with odd atomic numbers, also reflects different stabilities of atomic nuclei. The even-numbered elements form a lopsided 98 percent of metal abundances.

ISOTOPES GONE EXTINCT

In painting this picture of the fabrication of elements in stars, I have glossed over an important fact: stellar nucleosynthesis does not simply make elements, it makes specific *isotopes* of those elements. Most elements exist in several isotopic forms, though they typically consist dominantly of only one isotope. For example, carbon atoms exist as one of three isotopes: carbon-12, carbon-13, and carbon-14 (the numbers refer to the mass of each isotope, that is, its protons plus neutrons; all carbon atoms have six protons but different numbers of neutrons). However, carbon-12 accounts for most of the carbon in limestone, in the atmosphere, and in your body. Many isotopes are unstable and spontaneously break apart into more stable configurations; such unstable isotopes are said to be *radioactive,* the process of disintegration is *radioactive decay,* and the new isotope produced by the decay is *radiogenic.* In the case of the carbon isotopes above, radioactive carbon-14 decays to form radiogenic nitrogen-14, whereas carbon-12 and carbon-13 are stable.

Chondritic meteorites contain a veritable zoo of radiogenic isotopes, produced long ago when their radioactive (now extinct) parents rapidly decayed. What is particularly interesting about this observation is that the short-lived radioactive parents were made in supernova events and yet were apparently still alive when the meteorite formed in the early solar system. This implies that only a short time could have elapsed between the supernova explosion that created these isotopes and the solar system's formation. Perhaps the best example of such an isotope is radioactive aluminum-26, which decayed to magnesium-26 at a rate such that it was virtually all gone within a few million years after its creation. Yet the radiogenic magnesium-26 in meteorites resides inside aluminum-bearing crystals in meteorites, so it must have been live aluminum-26 at the time the crystals formed. In 1976, Typhoon Lee and Jerry Wasserburg at the California Institute of Technology measured the amount of magnesium-26 in chondrites and inferred that its live parent had to have been synthesized in a supernova less than 1 million years before the formation of our solar system, a remarkably brief interval. The former presence of other short-lived radioactive isotopes, such as chromium-53 and palladium-107, supports this startling conclusion.

These are not the only extinct isotopes in chondrites. Evidence for the former presence of iodine-129 and plutonium-244, both of which decay about ten times more slowly than aluminum-26, has also been found in these meteorites. These must have been fabricated in even earlier supernova events, perhaps 100 million years before the solar system formed. Thus, the incorporation of several generations of short-lived isotopes in meteorites suggests that at least two waves of supernova matter washed over our part of the universe at different times.

Hubert Reeves of the Centre National de la Recherche Scientifique in Paris has suggested a plausible explanation for these observations. Our star is just one of billions that comprise the Milky Way galaxy. Our position within the plane of this galactic disk can be discerned by observing the glittering band that the Milky Way cuts across the dark night sky. What cannot be so readily observed from our location is that this galaxy is an elegantly curved pinwheel. The spiraled arms are areas where the concentration of stars is particularly dense, and so it seems reasonable to assume that supernova explosions might be more common there. As our solar system spins about the galactic center, every 120 million years or so it passes though one of the spiral arms. The input of supernova-formed iodine and plutonium isotopes may have occurred during one such pass, and the addition of shorter-lived aluminum, chromium, and palladium isotopes happened on the next passage 100 million years later. Of course, both these passes happened before there was a solar system; only a cloud of gas and dust, from which the solar system would eventually be made, was actually salted with these radioactive isotopes.

RECOGNIZING STARDUST

Until the last few decades, it was generally believed that all the solid matter that went into the construction of our solar system had been vaporized and recondensed, thereby thoroughly and irretrievably altering it beyond recognition. However, peculiarities in the isotopic composition, that is, the mix of isotopes, of neon and xenon (both "noble" gases, since they do not combine with other, less exalted elements) in chondritic meteorites hinted that some exotic gas carriers might be present in them. In order to preserve anomalous gaseous isotopes in recognizable form, the noble gas carriers would have to be solid grains. David Black, then at the University of Minnesota, first suggested that tiny interstellar grains might be the carriers of certain isotopes of the odd noble gases in meteorites, and that eventually it might be possible to separate them from the much more abundant solar system matter that comprised most of the meteorites.

Confirmation of this idea required stubborn persistence and was a long time in coming. Chemist Edward Anders and his colleagues at the University of Chicago developed a procedure in which chondrites were sequentially attacked by acids, and at each step the residual material had to be checked to see if it still contained the carriers of neon and xenon isotopes. This is something like trying to retrieve salt grains in soup by evaporating the soup and then progressively removing the other solid materials in the residue, tasting for salt at each step. By 1987, after a decade of false leads and frustration, Anders finally succeeded in isolating a minuscule amount of white powder, the carrier of isotopically anomalous xenon. To everyone's amazement and delight, this material turned out to be miniature diamonds, crystals so small that they contained only a few thousand carbon atoms at most.

Buoyed by this success, the Chicago group soon separated tiny grains of another carrier, this time of anomalous neon isotopes. By joining forces with a research team at Washington University in St. Louis, they brought to bear new analytical techniques developed for characterizing extremely small specimens. The neon carrier was identified as silicon

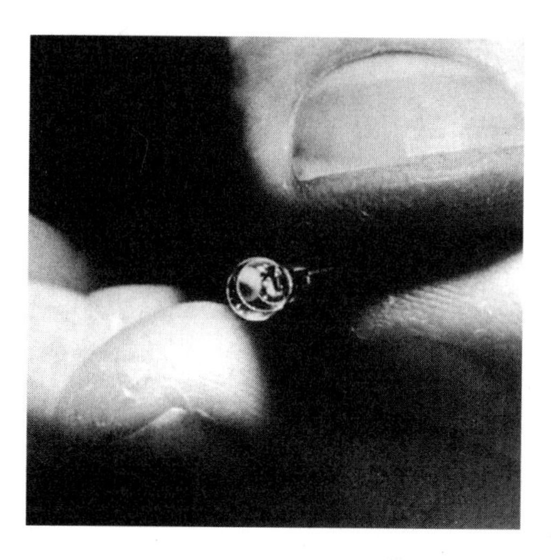

The white powder in the bottom of this tiny vial is a collection of minute diamonds, stardust formed from clouds of carbon sloughed off the surface of a red giant long before the solar system formed. These diamonds also contain implanted xenon, formed during a supernova that possibly destroyed this ancient star. The stardust was extracted from a chondritic meteorite. (Photograph by R. Lewis, University of Chicago)

carbide (used as an abrasive on Earth, and marketed under the trade name Carborundum). Ernst Zinner of the Washington University team measured the isotopic compositions of carbon and silicon in the tiny Carborundum particles, and their large isotopic differences from normal solar system materials proved conclusively that these were interstellar grains. More discoveries soon came rapidly. In 1990 the University of Chicago group separated yet another exotic carrier of neon, this time tiny spheres of graphite.

Diamond, silicon carbide, and graphite, which are all compounds of carbon, surely must have been coughed up from stars that had already synthesized a lot of this element. These particles may have formed from gaseous envelopes sloughed off of distended red giants, stars so large that their gravity could no longer hold on to their surface layers. The unusual isotopes of neon and xenon probably formed during later supernova explosions, and these newly synthesized atoms were propelled rapidly outward by the force of the blast. As they overtook crystalline clouds of diamond, Carborundum, or graphite, some of the noble gas atoms were implanted into the solid grains, forever tagging them as stardust.

Not all interstellar grains are compounds of carbon. Astronomers have ascertained from measurements of the spectra of interstellar dust in

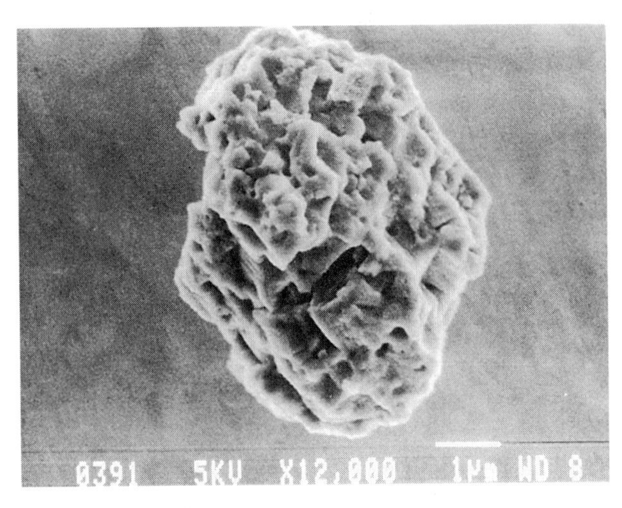

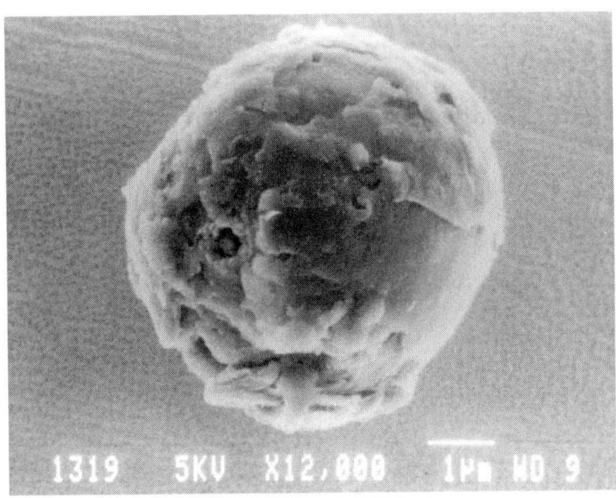

These images show minute specs of interstellar dust magnified thousands of times. The top grain is silicon carbide and the bottom is graphite, both extracted from a chondritic meteorite. The photographs were taken using an electron microscope, and the small scale bars in each image are 1 micrometer long. (Photographs by S. Amari, University of Chicago)

space that about half of such grains must be oxides or silicates (compounds of metals combined with oxygen, or with silicon and oxygen, respectively). Interstellar grains with these compositions are much more difficult to recognize in chondrites, because the bulk of the host meteorite consists of similar minerals formed within our own solar system. Nevertheless, after painstaking searches several workers have now identified the first interstellar grains of aluminum oxide, the mineral corundum, whose colored forms are ruby and sapphire. (One might get the

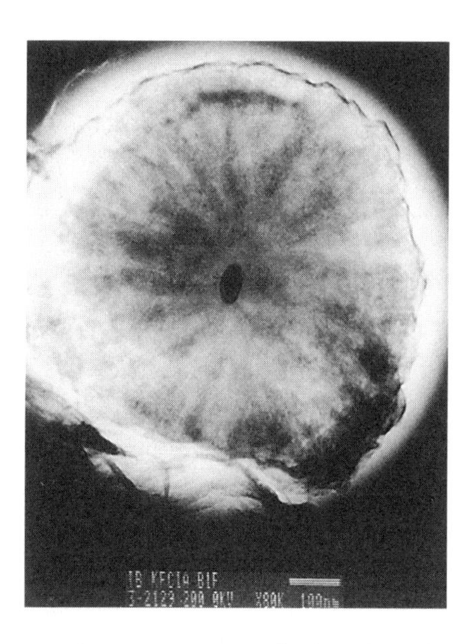

A transmission electron microscope image of a thin slice of an interstellar graphite grain reveals a hidden kernel of black titanium carbide at its center. The scale bar is 0.1 micrometer long. (Photograph by T. Bernatowicz, Washington University, St. Louis)

impression that the cosmos is a gem factory!) More recently, a tiny particle of titanium carbide has been found locked inside an interstellar grain of graphite, in effect an interstellar grain within another grain. Interstellar silicates have not yet been recognized, but this will be challenging because chondrites are mostly silicate minerals.

One other possible form of stardust has been suggested, but this one is less assured. Fullerenes are very large molecules, usually containing sixty carbon atoms arranged into a soccer ball–shaped cage. These molecules, nicknamed "buckyballs," are the rarest form of naturally occurring carbon on the Earth, much less abundant than even diamonds. Fullerenes were accidentally discovered in 1985 when scientists heated carbon vapor above 14,000 degrees. They are also thought to form in red giant stars. Recently, fullerenes have been found at the two largest impact craters on Earth, the Chicxulub crater in Mexico and the Sudbury crater in Canada. It has been suggested that these peculiar molecules formed inside stars and were later brought to Earth within the meteors that formed these craters, although it is also possible that they were produced from other forms of carbon during impact.

REFRIGERATED DEUTERIUM

Many chondritic meteorites are enriched in deuterium, the heavy isotope of hydrogen. Nothing on Earth has such a high ratio of deuterium to hydrogen (the chemist's notation for this ratio is D/H), so this enhancement cannot be due simply to terrestrial contamination. We have already noted that lots of deuterium was produced during the Big Bang, but it tends to be destroyed during nucleosynthesis in stars. It is *between* the stars that deuterium really gets concentrated.

Molecules form in interstellar space by chemical reactions (which form bonds between existing atoms), not by nuclear reactions (which create new atoms). The space between stars is frigid but, even so, reactions occur between charged gas atoms. Under such conditions, deuterium is preferentially used to construct the molecules, so that they end up with elevated D/H. In fact, the D/H in interstellar molecules is about a thousand times that of our solar system. What are these refrigerated, deuterium-enriched molecules? Radioastronomers have discovered that many are combinations of carbon with hydrogen, oxygen, nitrogen, and sometimes sulfur, molecules that a chemist would call *organic*. (The term "organic" does not imply any relationship to organisms, although living things are made of and produce organic matter.) Interstellar molecules are rather simple organic molecules though, normally just one or two carbon atoms with a few appendages of other elements.

The deuterium in chondrites is not spread evenly throughout, but is instead concentrated in organic matter. The organic compounds in meteorites, consisting of long chains of carbon atoms with complicated branchings, are much more complex than those observed in interstellar space. The complex forms must have been crafted inside the solar system from simpler interstellar molecules, the only known source of such enriched deuterium.

These findings have important implications for understanding the original stuff from which our solar system was made. Not only stardust, but also refrigerated molecules in clouds of interstellar gas must have been important constituents in the recipe for our world and its surroundings. In such clouds of gas, it is plausible that irradiation by cosmic rays could also have created the small amounts of lithium, beryllium, and boron that occur in meteorites.

A humorous but sobering commentary on our limited ability to appreciate the magic around us. (*Hagar the Horrible*, reprinted with special permission of King Features Syndicate)

ONE WITH THE COSMOS

Twinkling stars are really chemical factories whose assembly lines turn out elements heavier than hydrogen. The newly fabricated heavy elements are added to the hydrogen and helium that comprise the bulk of evolving stars. Stellar explosions periodically pollute the surrounding interstellar medium with their chemical products, where they are eventually recycled into new stars. The elements, like antiques, are used over and over again.

About 2 percent of the mass of our own Sun consists of heavy elements, so it cannot be a star of the first generation. Near the beginning of time, immediately following the Big Bang, the universe must have been chemically simple, almost pure hydrogen and helium. Stars made directly from this material contained virtually no metals. As best we can tell, our Sun may be a third-generation star (called a "Population I" star by astronomers). We can observe stars of the previous generation ("Population II"), which have a factor of a hundred or so lower abundance of heavy elements. However, only Population II stars with low mass survive to the present day. The more massive members of this group would have evolved faster, so that they no longer exist. A few Population III stars, from the first generation, might still exist, but they have not been observed in our galaxy. To last this long, they would have to be very small and thus would be very dim.

Our solar system was constructed in part from refrigerated organic molecules and cosmic ray–produced light elements, the matter of cold clouds of gas between the stars. As the cloud that was destined to become our Sun rotated through the galactic spiral arms, it was seasoned, more than once, with heavy elements newly forged in ruddy stellar giants.

Almost all of the unimaginably vast assortment of matter from which our solar system was crafted has been molded into other forms, most now unrecognizable as primal. The only exceptions, so far as we know, are found in chondritic meteorites. Within chunks of such meteorites are the twisted remnants of interstellar organic molecules, tarry gunk whose interstellar ancestry is fingerprinted in its deuterium enrichment. Sometimes surviving in more pristine form are tiny interstellar grains of carbon or oxide, microscopic shards older than the Sun itself. These crystals are recognizable only because they have peculiar isotopic abundances of carbon or silicon, or are tagged with supernova-generated noble gases, testifying to their incredible histories. Judging from the isotopic variations so far measured in interstellar grains, the gas cloud from which our own solar system formed must have received additions of matter from perhaps a hundred other previously existing stars. The chondrites are thus revealed to be lithic sanctuaries of cosmic memory.

This assortment of starry ash and organic muck, this interstellar dross, seems a paltry inheritance from the universe, but it affords a unique opportunity to hold in our hands samples of the ancient matter from which the solar system formed. Moreover, these tiny motes provide elegant confirmation of a truly breathtaking idea: that every atom of silicon or iron that comprises our beautiful world, every calcium atom in our bones or carbon atom in our bodies, was made in a star. We, too, are stardust and antique elements, the immortal stuff of the cosmos, its recycled past and, in all probability, its future.

Some Suggested Readings

Allegre, Claude (1992). *From Stone to Star: A View of Modern Geology.* Harvard University Press, Cambridge, MA. *This superbly written book has an especially good and complete explanation of nucleosynthesis in stars.*

Bartusiak, Marcia (1993). *Through a Universe Darkly.* HarperCollins, New York. *Chapter 5 of this excellent, nontechnical book describes the history of discoveries about the alchemy of stars in rich detail.*

Kirschner, R. P. (1994). The Earth's elements. *Scientific American,* vol. 271, no. 4, pp. 59–65. *A particularly lucid account of how the elements are produced in stars, as well as a description of supernova observations.*

McSween, H. Y. Jr. (1993). *Stardust to Planets: A Geological Tour of the Solar System.* St. Martin's, New York. *The fifth chapter of this book presents the detective story of the discovery of stardust in meteorites, and the fourteenth describes how meteorites can be used as proxies for cosmic abundances.*

Parker, Barry (1993). *The Vindication of the Big Bang: Breakthroughs and Bar-*

riers. Plenum, New York. *A recent history of cosmology, focusing on the Big Bang model.*

Peebles, P. J. E., Schramm, D. N., Turner, E. L., and Kron, R. G. (1994). The evolution of the universe. *Scientific American,* vol. 271, no. 4, pp. 53–57. *The Big Bang is described in this article, along with the subsequent expansion of the universe.*

Play It Again, Johannes

Nebular Accretion of the Earth and Planets

HARMONY

It would be a stretch to describe playing trombone in a high school marching band as a profound musical experience. At the same time I was strutting to Sousa, however, I spent some of my weekends as a member of a dance orchestra, playing the scores of Glenn Miller, Stan Kenton, and other bands of the swing era, and that interlude shaped my musical tastes even to this day. This anachronistic band was understandably more in demand by people of my parents' generation than by my fellow students, but the experience was nonetheless rewarding in that it gave me a greater appreciation for harmony than did the rock-and-roll staple of my own generation. In other forms of music like classical or jazz, I find that I am still moved more by well-crafted chords and chord progressions than by lilting melody, technical virtuosity, or clever improvisation.

As much as I am drawn to harmony, my appreciation of it must pale in comparison to Johannes Kepler's. Born in 1571, Kepler suffered an attack of smallpox as a child. This illness left him with crippled hands and damaged eyesight, which squelched any musical aspirations he might have had but not his love of harmony. In school Kepler's musical interests were supplanted by math and science, for which he showed considerable aptitude, and in the course of his studies he became an early convert to the heliocentric solar system of Copernicus. At the age of twenty-four, Kepler published a book defending the Copernican system but also espousing a peculiar idea that was his very own. He tried to fit

Copernicus' circular planetary orbits inside gigantic geometric figures that were based on the five perfect solids: the octahedron, icosahedron, dodecahedron, tetrahedron, and cube. (The "perfection" of these solids stems from the fact that these are the only forms bounded by faces that are identical regular polygons.) First, he fitted an octahedron around the orbit of Mercury, and found that a sphere circumscribed around the octahedron corresponded to the orbit of Venus. Next, he noticed that a sphere circumscribed around an icosahedron he had fitted around the orbit of Venus corresponded to the orbit of the Earth. Continuing the process outward and using the remaining perfect solids, he described geometrically the orbits of all six of the then-known planets—Mercury, Venus, Earth, Mars, Jupiter, and Saturn. An added bonus in Kepler's geometric scheme was that it neatly explained why there were only six planets: there could be no more planets than there were perfect geometric solids.

Kepler was delighted with his odd discovery, but it failed to arouse similar enthusiasm in others. He sent along a copy of his book to Tycho Brahe, indisputably the world's greatest naked-eye astronomer, who criticized it severely. Nevertheless, Tycho must have been impressed enough with Kepler's mathematical abilities, for he soon issued an invitation to the young scholar to join him as his assistant in Prague, then a mecca for science and culture. The irascible Tycho was certainly not an easy or even likable employer, but Kepler could ill afford to pass up an opportunity to work with twenty years' worth of Tycho's very precise observations of planetary motions (and besides, he had just lost his job as a teacher during a religious purge and had been given until sundown to get out of town). Tycho needed Kepler's mathematical skills to make sense of his gold mine of data, but he was so jealous of Kepler's expertise that he doled out his observations rather slowly. Only after Tycho died suddenly, thus making the records available in their entirety, did Kepler make any substantial progress in interpreting these observations.

A man of few social graces himself, Kepler has been described as slovenly and awkward. Whatever his faults, he could not be criticized for being lazy. Launching into this work, Kepler initially tried to fit Tycho's data to circular planetary orbits, as had generations of astronomers before him. Eventually he was forced to try less traditional shapes. Kepler's success came only after he abandoned the view that planets must move in perfect circles, an age-old tenet that had imprisoned astronomy for centuries. After many false starts and years of mind-numbing calculations he finally succeeded, publishing in 1609 his first two laws of

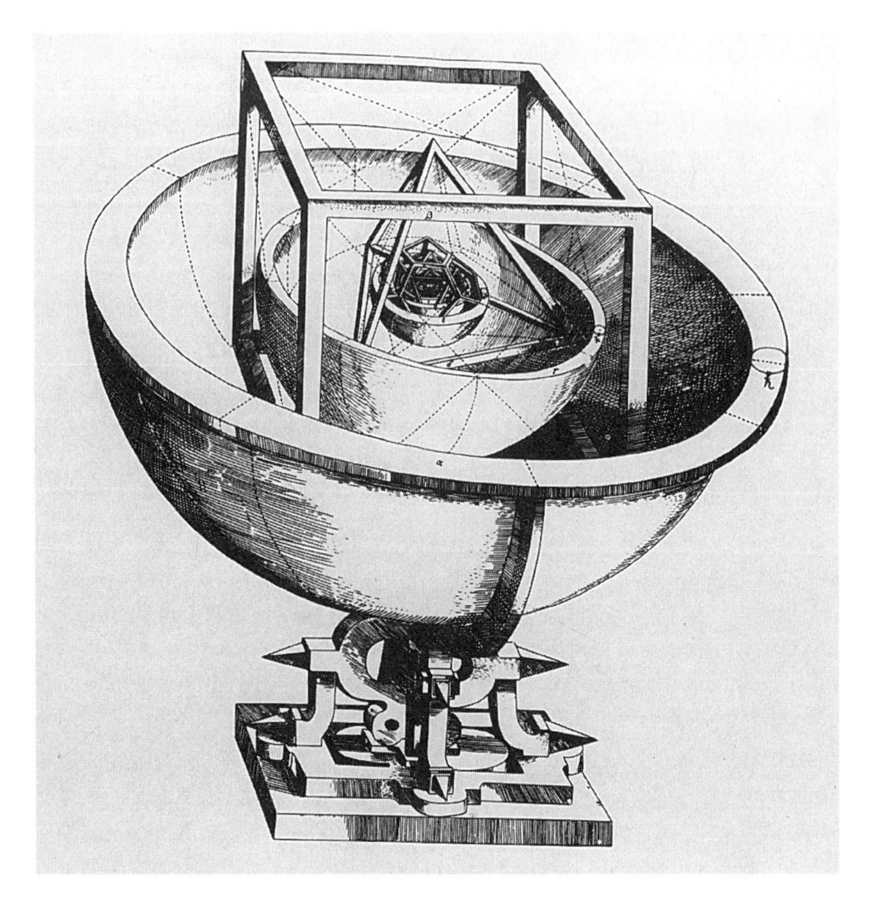

A peculiar drawing by Johannes Kepler illustrating his geometrical fancy that each planet's orbit could be enclosed by a perfect solid and an encasing sphere. The outermost sphere and cube represent the orbit of Saturn; the orbit of Mars is represented by a smaller sphere enclosing a tetrahedron; and an even smaller sphere enclosing a dodecahedron represents Earth's orbit. (From Kepler's *Mysterium Cosmographicum*, published in 1596)

planetary motion. In this work he offered the radical proposal that planets follow elliptical paths.

At about that juncture, the musician in Kepler began to assert itself. This was to prove as unfortunate a diversion as had his earlier foray into geometry. He had always been intrigued with the notion of harmony in nature, but not in the way that we might use this term; Kepler interpreted the concept of harmony in nature quite literally. In his third law, published in 1618, he uncovered a simple mathematical relationship between the distance of a planet from the Sun and the time it takes to complete a revolution. I suppose that you could say that this relationship does describe a sort of harmony in the motions of planets, but Kepler carried this idea to absurdity; he imagined that the planets sang out notes of music as they moved in their orbits, at least in a figurative sense. Using the established relationship between the lengths of strings and the musical tones they produced when strummed, Kepler converted the dimensions of planetary orbits into musical tones. He envisioned that each planet resonated one note at a time, so that together the planets hummed in harmony. Although this music could only be heard by the Creator, Kepler filled his *Harmony of the Worlds,* the book describing his third law, with staves of music.

THE GRAVITY OF THE SITUATION

In answering one question about the shapes of planet orbits, Kepler had revealed yet another: Why did the planets move in ellipses? In the classical view, the circle was nature's own curve, perfection as befits the Creator. But if nature had repudiated the circle, another kind of answer must be called for. Kepler wrestled with this problem with little success. Its answer was forthcoming only later with Isaac Newton's explanation of universal *gravity.*

As early as 1666, Newton had understood that the force that kept planets swinging about the Sun in Keplerian orbits was gravity, and that their peculiar elliptical shapes were an expected consequence of the way that this force varies with the distance between each planet and the Sun. He might have published this groundbreaking insight immediately, but he wanted to go further and show that the same mysterious force that regulated the motions of planets also caused an earthly apple to fall to the ground. Not until eighteen years later did Newton's professional colleagues prevail upon him to submit his discovery to the Royal Society for publication.

Nearly a century passed before Newton's concept of gravity actually began to shape ideas about how the solar system formed. Catastrophic theories are now out of favor, but for a time they were very popular. Georges-Louis Buffon appears to have been the first, in 1778, to suggest that the planets formed when a comet crashed into the Sun. In this scenario, a plume of matter that was splashed outward during the collision remained gravitationally bound to the Sun and eventually was assembled into planets. In those days comets were thought to be similar to stars, and when the fact later emerged that comets are much too small to have any appreciable effect on the Sun, the colliding comet was replaced by a passing star whose gravity pulled Sun stuff off like taffy, matter that later formed planets.

POLYPHONY

Polyphony is a term referring to the musical juxtaposition of two or more independent themes, usually with a pleasing, harmonious result. Newton himself had been intrigued with the observation that the planets all lie approximately in the same orbital plane as does the Earth, the so-called *ecliptic,* because his law of gravitation did not require that. The polyphony of combining this critical observation with Newtonian gravity provided the key to unraveling the true origin of the planets, but Newton was not the one who took this step. Rather, it was Pierre-Simon Laplace who, more than a century later, melded gravity with the planar solar system.

As an instructor at the École Militaire (where he taught Napoleon Bonaparte), Laplace actively engaged in scientific research, and over a seventeen-year period he published an immense number of important papers. Not satisfied to shine only in scientific circles, he also attempted a political career following the French Revolution, only to be summarily fired after six weeks as a rather incompetent minister of the interior. But the government's loss was science's gain. In 1796, Laplace published his "nebular hypothesis" that the solar system formed from a collapsing mass of gaseous matter.

Such a cloud, called the *nebula* from the Latin word for "fuzzy," would likely have some intrinsic circular motion, or angular momentum (the latter is actually defined as velocity times mass). As this gaseous material cooled it shrank, and Newtonian gravity required that it rotate faster and faster to conserve its angular momentum. The same principle of physics is employed to good advantage by figure skaters. I confess that I cannot recognize the difference between a double axel and a triple lutz without the guidance of a commentator, but I find the technique and physics of a spinning ice skater astonishing. In no other arena of sport can a human being move so fast as to become a blur. A skater's spin begins slowly, but as she pulls her arms into her body she rotates noticeably faster, even though no further force is imparted to her skates. The skater's angular momentum is conserved, in accord with Newton's law of gravity, and the crowd applauds.

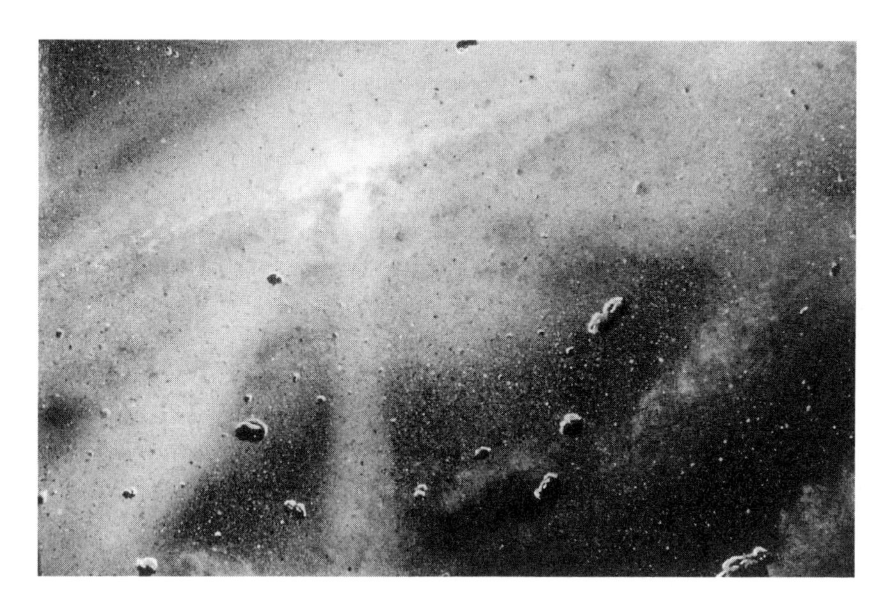

An artist's conception of the solar nebula illustrates its disklike shape. This painting, by planetary scientist William Hartmann (Planetary Sciences Institute), emphasizes that the nebula was composed of matter that soon accreted into small, rocky planetesimals.

In Laplace's nebular hypothesis, eventually rotation caused the nebular gas to flatten into a disk with a bulge in the center. (If figure skaters could rotate fast enough to bulge perceptibly in their middles, the spin might never have become part of this competition in grace.) The formation of a nebular disk was a particularly important aspect of Laplace's model, because it provided an explanation of why all the planets lie in the ecliptic plane. Once formed, the disk fragmented as rings of matter were spun off in sequence, with the material in each ring destined to be gravitationally assembled into a planet. As each planetary kernel of gas cooled and shrank, it too spun faster and sometimes shed rings of material that ultimately formed satellites.

A thorny problem for Laplace's nebular hypothesis, then and now, is that the present Sun should be rapidly whirling like a figure skater. Instead, this star, with 99.9 percent of the solar system's mass, is spinning much too slowly, having only 2 percent of the system's angular momentum. After grappling unsuccessfully with this problem, many physicists returned to catastrophic theories in the early part of this century. However, it was soon demonstrated that a filament of Sun dislodged by a passing star would disperse into space before it could collapse into planets. The statistical improbability of two stars colliding, as well as other problems with this idea, finally led to its shelving. The nebular hypothesis has again become the almost universally accepted model, and novel ways of solving the Sun's angular momentum problem have been proposed.

Formation of planets within a nebula setting could conceivably occur by either of two rather different mechanisms. Laplace envisioned the planets forming by *condensation,* that is, the transformation of gas into solid substance. At some early stage just before condensing in this way, the Earth and the other planets would thus have been giant gas balls, sometimes called gaseous protoplanets. This model, popular at the midpoint of this century, requires a massive nebula, with perhaps twice the mass of the Sun, to induce it to fragment into separate bodies. That seems implausible, however; planets only have about a thousandth of the mass of the Sun, implying that the mass of the nebula at the time the planets began forming was also small. The alternative is that planets formed gradually by *accretion,* the gradual assembly of small pieces of condensed matter into larger and larger objects.

DISKS AND STELLAR INFANTS

Real progress in understanding how our solar system formed was virtually nonexistent from the time of Laplace until just a few decades ago, but advances are now moving at a dizzying pace. Both observation, using telescopes with breathtaking resolution, and theory, made possible by unprecedented computational power, have spurred this progress.

The formation of planets like our own is merely incidental to star formation, a process on which modern astronomy has shed considerable light. Most of the cosmos is incredibly empty, containing only one atom in each cubic centimeter of space. However, here and there among the stars are dark molecular clouds, regions where matter is concentrated a 100- or 1,000-fold. Molecular clouds were first discovered by the great eighteenth-century astronomer William Herschel, although he did not recognize them for what they were. As he counted stars in different regions of the sky, Herschel could not help but notice inky splotches in the curtain of stars, regions that appeared to be devoid of matter. On one night he exclaimed, "Here is truly a hole in the heavens!" Not until more than a century later was it finally accepted that these haunting, black regions were actually dense accumulations of dark matter that blocked the starlight behind them.

This molecular cloud, made of concentrated gas and dust, is called the Eagle Nebula. A Hubble Space Telescope image reveals it to be a large nest of newly hatching stars, each embedded into the top of a fingerlike protrusion. The embryonic stars protect the gas and dust behind them from erosion by ultraviolet light from other nearby hot stars, creating the fingers. (NASA)

As their name suggests, most of the gas atoms in *molecular clouds* are combined into molecules, such as deadly carbon monoxide and methane, smelly ammonia and hydrogen sulfide, and water; even ethyl alcohol has been found in staggering quantities, though spread so thinly through the clouds that an area as big as Jupiter would have to be distilled to make a cocktail. Molecular clouds are really chemical crucibles, converting the elements created in stellar nuclear reactors into molecules. However, dust particles composed of silicates, sometimes coated with ice, have also been observed. The dust, for the most part, is an inheritance from supernova explosions. Some molecular clouds are huge, with enough mass to make 100,000 suns. The orbiting Hubble Space Telescope has imaged several molecular clouds—hulking maelstroms located thousands of light years distant from Earth.

A Hubble Space Telescope image of the star Beta Pictoris shows a vast dust disk, viewed edge-on. The photograph was taken with a technique that blots out light from the bright star itself, which is located at the lower left corner. (NASA)

It is within these dense, cold nurseries that stars are incubated. Shifting masses of gas periodically form clumps that are gravitationally unstable, collapsing chaotically into stellar conflagrations. Temperatures and pressures in such collapsing regions rapidly reach enormous levels, tens of millions of degrees and billions of atmospheres, triggering the fusion of hydrogen. Like tadpoles, stars are born hundreds at a time, often remaining clustered together for a while after their formation. As a result of communal hatching, many stars, perhaps most, are twins or even triplets, multiple stars that are so close that they orbit about each other. Recent surveys suggest that perhaps only a tenth of all stars are solitary. For reasons not fully understood, molecular clouds sometimes fragment, and our lone star is thought to have formed from the gravitational collapse of such a fragmented mass of gas and dust.

Stellar astronomers have a decided advantage over planetary scientists, in that they can look backward in time. The universe is immense enough that astronomers can find and observe stars at various stages of their formation and evolution. When these snapshots of stellar history are strung together, they can serve as an accelerated movie of our Sun's creation and evolution, rich with details about which we could otherwise only speculate.

Laplace guessed that the luminous blurs discovered by the astronomers of his day were nebulae in various stages of collapse. Later observations with higher-powered telescopes, however, showed that these blurs were actually entire galaxies. Nevertheless, as envisioned by Laplace, molecular cloud fragments really do rotate and flatten into nebular disks. Modern astronomers have demonstrated that many young stars are surrounded by disk-shaped nebulae similar in dimension to our own solar system. Some of these disks are luminous, implying that gas and dust are actively being plastered onto the protostar, so that its central mass will soon become a sun. A visual, close-up image of a 600-million-kilometer-thick disk around one nearby star, Beta Pictoris, illustrates its form. This striking photograph, of a disk viewed nearly edge on, would have greatly excited Laplace, as it does modern astronomers, but it is probably not really a nebula. Beta Pictoris, although thought to be billions of years younger than our Sun, is still no infant star, and its disk is probably only a surviving relic of a nebula that formerly enclosed it. The shape of this disk is better described as a donut than a pancake, and invisible planets may already orbit within a cleared area inside the visible disk. Other disks associated with fledgling stars are darker but have been imaged at nonvisible wavelengths (they glow in the infrared and millimeter wavelengths). The Hubble Space Telescope has imaged no less than 150 visible circumstellar disks in the Orion Nebula, revealed as a glowing nest of gas studded with brilliant infant stars.

As well as demonstrating that nebulae exist, astronomical observations also tell us something about how they collapse. Within the constellation Taurus is a star named T Tauri, a stellar infant with about the same mass as the Sun. Immature stars like T Tauri have not yet settled onto the main sequence where stars spend most of their existence, but rather they radiate energy at a furious pace. Instead of quietly munching infalling dust and gas, these infant stars voraciously gobble up the matter around them. Another particularly violent young star, FU Orionis, exhibits wild oscillations in brightness, apparently tantrums of conspicuous consumption powered by accretion of matter from the disk in fits and starts. FU Orionis also spits out matter in ferocious winds that emanate from its rotational poles.

To learn more about nebula collapse, we must turn to theoretical

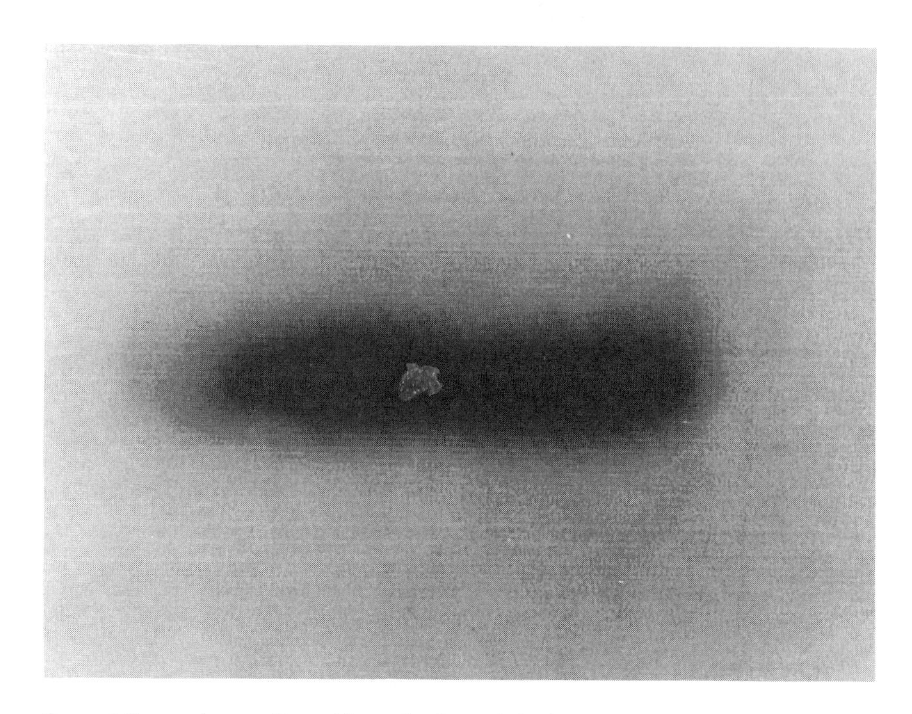

Resembling an interstellar Frisbee, this hazy disk-shaped nebula, seen edge-on, was imaged by the Hubble Space Telescope. A fledgling star is hidden inside the large disk (seventeen times the diameter of our own solar system), its light blocked by dust. This planetary system under construction is situated in the Orion Nebula, fifteen hundred light years away. (NASA)

models. Because of the complexity of the mathematical equations that describe how turbulent clouds of gas collapse, these models are best handled as computer simulations. One conclusion from such studies is that collapse of our own nebular disk happened from the inside out, meaning that matter close to the center was added to the growing Sun faster than matter farther out could take its place. When material in the rotating disk is not distributed symmetrically about its axis, gravity tugs on some regions more strongly than on others, causing an imbalance that moves matter inward. This mechanism, called gravitational torque, also would have induced the outer part of the nebula to rotate faster, just as an automobile engine delivers torque to the driveshaft.

Molecular clouds can begin to collapse on their own, as especially dense areas feel their own gravitation. In the case of our own solar system, though, there is evidence of some external triggering event. The former existence of certain short-lived radioactive isotopes (especially aluminum-26) in chondritic meteorites implies that less than 1 million years could have elapsed between their synthesis in stars and their incorporation into the solar nebula. Such isotopes were probably injected into interstellar space during a supernova event, but this short time interval suggests the stellar explosion must have occurred just next door. Shock waves from the explosion may have compressed the gas ahead of the expanding shock front, thereby triggering its collapse into a nebula. Our own solar system may owe its birth to the death of a neighboring star.

TOO HOT TO HANDLE

When you inflate a tire with a bicycle pump, the barrel of the pump gets hot as the air in the chamber is compressed. In an analogous way, collapse of the solar nebula compressed and heated the gas from an initial temperature of −260 degrees to perhaps 1,200 degrees Celsius in the inner planet region. Sudden outbursts of energy as matter accreted to the growing Sun also must have resulted in dramatic temperature increases. Nebular temperatures were strongly dependent on whether the nebula was opaque, because an opaque disk cannot radiate heat away. The outer reaches of the nebula, where the density of dust was lower and opacity was less, were much cooler than the inner nebula. The effect of this heating, at least in the inner nebula, was to fry the rotating interstellar dust, causing much of it to melt or even to vaporize. We might predict that the only surviving matter from the inner solar system would be *refractory* in composition, nuggets of elements impervious to high temperatures and from which more *volatile* elements were boiled off. However, in the cooler outer reaches of the nebula, some tiny interstellar grains, like diamond and Carborundum, might have survived.

What happened to the matter that was vaporized? Some was certainly incorporated as gas into the Sun, but some may have later condensed, that is, reconverted back to solid form as it cooled. In the Earth's atmosphere, which is relatively dense compared to the nebula, water vapor condenses into liquid (raindrops) but not directly into solid ice (snowflakes are crystallized raindrops). However, at the much lower pressures in the nebula, gas condenses to form solids directly rather than liquids. The condensation idea has been around a long time and, in fact, was part of Laplace's original nebula hypothesis and even earlier catastrophic models in which the solid planets formed from gaseous matter pulled out of the Sun. In 1972, condensation was explored in depth by Larry

Grossman, then a graduate student at Yale. Beginning with a hot gas of solar composition, Grossman calculated the order in which various minerals should condense as the gas cooled. This was an undertaking especially suitable for computer manipulation, since he had to consider all possible reactions among gaseous and solid materials at different temperatures to determine which elements would condense and which would remain in the gas. He found that, at the highest temperatures (beginning about 1,400 degrees Celsius), the first condensates were refractory minerals (mostly oxides of calcium, aluminum, and titanium), the same kinds of compounds that are used to make the bricks that line industrial furnaces. Other, less refractory minerals formed with further cooling of the gas. Especially important steps in the condensation sequence occur at about 1,200 degrees when metallic iron condenses, followed in short order by large quantities of magnesium silicates.

A recent revision of the condensation sequence that attempts to model the conditions in the inner nebula region more closely, by Akihiko Hashimoto and John Wood of the Smithsonian Astrophysical Observatory, gives similar results, except that iron metal and magnesium silicates switch places in the sequence. Iron is a particularly abundant element, and correctly understanding its condensation behavior is critical. It has been suggested that the nebula temperature may have been regulated by the condensation of iron dust, acting like a thermostat. If the nebula became too hot, iron would vaporize and opacity would drop, causing cooling; if the nebula cooled to the point where iron condensed, the opacity would increase, thereby adjusting the temperature upward.

As nebular temperatures continued to drop below the iron condensation point, previously condensed solids reacted with the gas in the nebula to form other minerals. For example, iron metal reacted with sulfur gas to form iron sulfide. The calculations also suggest that water vapor reacted with silicates to produce hydrous solids like the terrestrial mineral serpentine. Eventually, whatever water was left in the gas finally condensed as ice. To make a long story short (if it is not too late already), the condensation sequence predicts the order in which solids appeared in the cooling solar nebula.

The Condensation Sequence

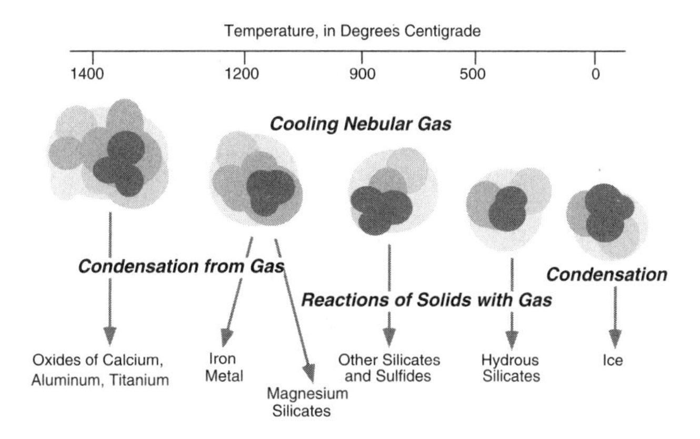

This schematic diagram illustrates the condensation sequence of solids from a cooling nebular gas of solar composition. Oxides of refractory elements like calcium, aluminum, and titanium form at the highest temperatures, followed by iron metal and magnesium silicates. At lower temperatures previously condensed minerals react with the gas to form other silicates and iron sulfide. Finally, at the lowest temperatures silicates react with water vapor to form water-bearing minerals, and ice condenses.

SAMPLING THE NEBULA

Sometime after midnight on February 8, 1969, a large, bright meteor entered the Earth's atmosphere and broke into thousands of pieces. This supersonic gravel bank plummeted to the ground and scattered over an elliptical area 50 miles long and 10 miles wide in the state of Chihuahua in Mexico. The next morning, the first of many meteorites from this fall was found only a few steps from a house in the village of Pueblito de Allende. Newspapers carried the story of the fireball, scientists raced to Mexico from the United States, and local children were let out of school to search for more specimens. Roughly two tons of meteorite fragments were recovered, all of which bear the name *Allende* for the location of the first discovery.

Individual specimens of Allende are covered with a black, glassy crust formed as their exteriors melted during atmospheric deceleration. When broken open, Allende stones are revealed to contain an assortment of small objects, spherical to irregular in shape, embedded in a dark gray matrix. These peculiar objects in Allende, and in other meteorites like it, were once constituents of the solar nebula.

The Allende meteorite is classified as a carbonaceous chondrite. Chondrites take their name from the Greek word *chondros* meaning "seed," an allusion to their appearance as rocks containing tiny grains. These seeds are actually *chondrules,* millimeter-sized melted droplets of silicate material that have been quenched into spheres of glass and crystal. A few chondrules contain relict grains that survived the melting event, so these enigmatic balls must have formed when clumps of nebular dust were fused at rather high temperatures approaching 1,700 degrees Celsius. Study of the textures of chondrules reveal that they cooled rather quickly, in times measured in minutes or hours, so the heating events that formed them must have been localized. It seems very unlikely that large portions of the nebula were heated to such extreme temperatures, and huge nebula areas could not possibly have lost heat so fast. Chondrules must have been melted in small pockets of the nebula that were able to lose heat rapidly. The origin of these peculiar glassy beads remains an enigma.

Equally perplexing constituents of Allende are the *refractory inclusions,* irregular white lumps that tend to be larger than chondrules. These aggregates are composed of minerals uncommon on Earth, all rich in calcium, aluminum, and titanium, the most refractory of the major elements in the nebula. The same minerals that occur in refractory inclusions are predicted to be the earliest-formed substances in the condensation sequence, an observation that initially caused quite a stir in the scientific community. However, studies of the textures of inclusions reveal that the order in which the minerals appeared differs from inclusion to inclusion, and often does not match the theoretical condensation sequence in detail. Some of the refractory inclusions thus may actually be residues left from the evaporation of more volatile constituents, probably with some recondensed matter added later.

Chondrules and inclusions in Allende are held together by a matrix of fine-grained, mostly silicate minerals, but also including grains of iron metal and iron sulfide. At one time it was thought that these matrix grains might be pristine nebular dust (one researcher termed this "holy smoke"), the sort of stuff from which chondrules and inclusions were made. However, detailed studies of chondrite matrix suggest that much of it, too, has been formed by condensation or melting in the nebula, although minute amounts of surviving interstellar dust are mixed with the processed materials.

All these diverse constituents are aggregated together to form chondritic meteorites, like Allende, that have chemical compositions much

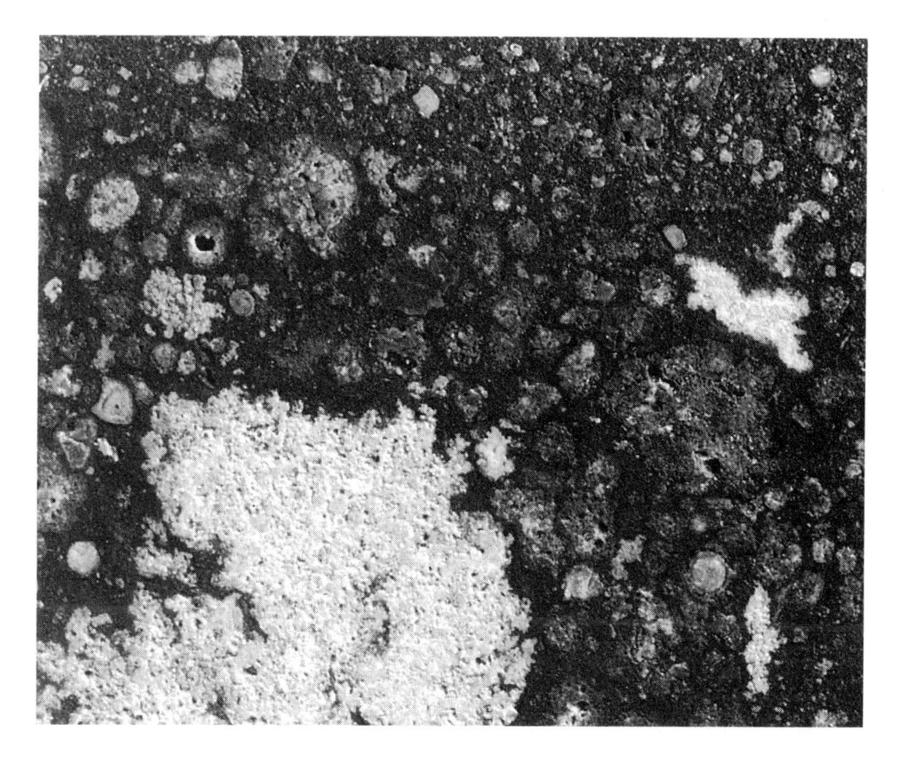

A slab of the Allende meteorite, measuring several centimeters across, shows abundant round chondrules and irregular (white) refractory inclusions, set in a dark gray matrix of fine-grained minerals. This chondritic meteorite is an aggregate of objects that once were suspended in the solar nebula. (Smithsonian Institution)

like that of the Sun. To compare the compositions of a meteorite and the Sun, it is necessary that we use ratios of elements rather than simply the abundances of atoms. After all, the Sun has many more atoms of any element, say iron, than does a meteorite specimen, but the ratios of iron to silicon in the two kinds of matter might be comparable. An accompanying figure compares the elemental composition of the Sun and Allende, with the number of atoms of each element given per 1 million silicon atoms. The compositional similarity is striking (elements falling along the diagonal line have precisely the same abundance ratios in the Sun and the meteorite). The major difference is that Allende is depleted in the most volatile elements, like hydrogen, carbon, oxygen, nitrogen, and the noble gases, relative to the Sun. These are the elements that tend to form gases even at very low temperatures. We might think of chondrites as samples of distilled sun, a sort of solar sludge from which only gases have been removed. Since practically all of the solar system's

Solar and Chondrite Compositions

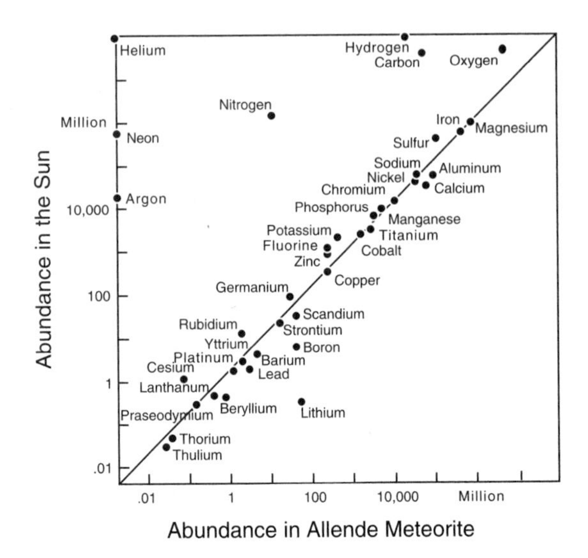

The chemical composition of the Sun, illustrated here in terms of ratios of the number of atoms of each element per 1 million atoms of silicon, is compared with that of the Allende meteorite. Elements having exactly the same abundance ratio in both fall along the diagonal line. Allende is chemically very similar to the Sun, except for its depletion in the most volatile elements like hydrogen, nitrogen, oxygen, carbon, and the noble gases.

mass resides in the Sun, this similarity in chemistry means that chondrites have average solar system composition, except for the most volatile elements; they are truly lumps of nebula matter, probably similar in composition to the building blocks from which planets were assembled.

Not all chondritic meteorites are carbonaceous like Allende; other chondrite types are similar in gross appearance and composition but have subtle chemical differences. These minor variations on the chondrite theme provide important insights into how planets formed. For example, volatile elements are depleted, relative to the Sun, in different chondrite types to varying degrees. This probably results from incomplete condensation at different locations and times in the nebula. The terrestrial planets, which probably formed closer to the Sun than did the meteorites, show the same kinds of volatile element depletions that chondrites do, only to greater extremes. Another variable among chondrites is the amount of iron metal they contain. This segregation of metal from silicates to varying degrees probably happened because metal grains are heavier than similarly sized silicate grains. Iron dust would have settled to the midplane of the nebula faster than silicate. Like the chondrites, the terrestrial (inner) planets contain different amounts of iron. Both of these chemical *fractionations,* that is, separations of elements from one another, support the idea that planets formed by accretion of already condensed solid matter.

The composition of the solar system is a grand average of all the assembled interstellar gas and dust. Chondrites like Allende provide the best estimate of this composition, because we can analyze them more precisely than the Sun. Allende also shows us that most of the matter in the nebula did not remain in its original form, but was processed in the nebula by strong, localized heating events that caused it to melt or vaporize, and ultimately to condense again in solid form.

THE PLANETARY CRESCENDO

The collapse of the solar nebula was rapid, complete perhaps within a few million years. We have already seen how fledgling stars display tantrums—explosive bursts of energy—and there is no reason to think that our infant Sun acted any better. This violent period may not seem to be an auspicious time to grow planets, but at this point the rotating disk of micrometer-sized particles and millimeter-sized melted clumps of dust was assembled into objects the size of marbles, basketballs, automobiles, and eventually kilometer-sized bodies called *planetesimals.* Exactly why these bits of matter began to stick together like Velcro is unclear; perhaps tiny particles were attracted to each other by surface charges, and dust adhered to the surfaces of molten chondrules.

Unlike the formative stages of stars, the act of planet formation has never been observed. Before about 1940, scientists accepted that concentrations of condensed nebular dust would simply contract into planets because of gravity, in much the same manner that the Sun formed. However, a new school of thought appeared in the Soviet Union, isolated from the rest of the scientific community by the Iron Curtain, at about that time. Pioneering astrophysicist O. Schmidt and his students Boris Levin and Victor Safronov showed that once these clumps reached planetesimal size, gravity took over and gentle clumping became intemperate smashing as these larger objects assembled themselves into bodies of even greater size. When colliding objects have different sizes, the smaller one tends to adhere to the larger one; this is exactly what happens when a meteorite falls to Earth. In this way, planetesimals attained ever

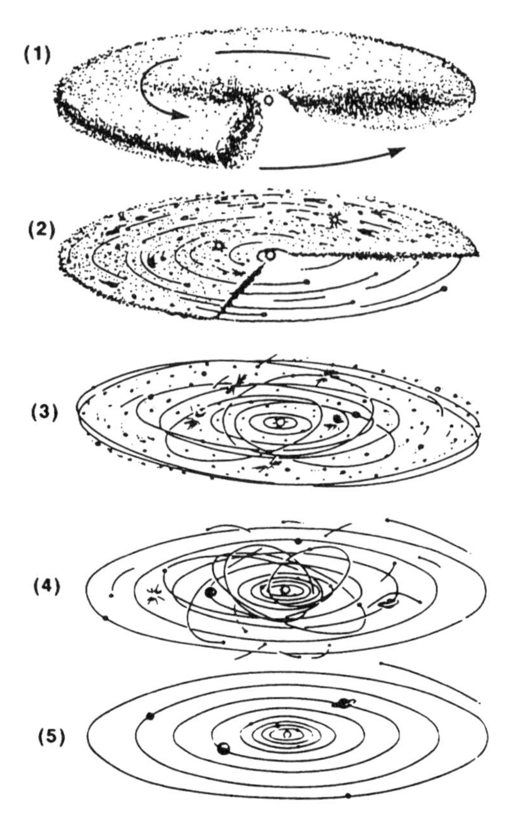

A series of sketches, by Russian astrophysicist Boris Levin, illustrate five progressive stages in the accretion of planets, from an initial dusty nebula, through the planetesimal stage, and finally to planets.

larger sizes. The most massive of the planetesimals grew at faster rates than their diminutive neighbors, a process referred to as runaway accretion. Computer simulations of the inner planet region suggest the former presence of at least a hundred objects with the mass of the Moon, ten objects similar in mass to Mercury, and several objects exceeding the mass of Mars. Such a swarm was the final population of planetesimals, the true building blocks of the Earth and its cosmic neighbors.

Each planet was once thought to have had its own *feeding zone,* an area straddling the planet's present orbit that was swept clean during accretion. In this scenario, each feeding zone had a particular chemical composition (because different parts of the nebula experienced different amounts of heating and condensation), leading to differences in the compositions of the planets that accreted within it. However, George Wetherill of the Carnegie Institution of Washington has questioned the validity of this concept. His computer simulations of accretion indicate that all the inner, terrestrial planets received contributions from at least as far out as the present asteroid belt. The differences in planetary compositions apparently cannot be explained simply by distinct feeding zones.

Earlier we saw that condensation theory predicts that water first condensed when the nebula cooled below 250 degrees Celsius, as water vapor reacted with dry silicates to form hydrous minerals like serpentine. However, the formation of water-bearing silicates in the nebula would have been ponderously slow, possibly requiring more time than the nebula existed. It seems more likely that most of the water would have condensed directly as ice when nebula temperatures became sufficiently low (below zero). Water is thought to have condensed as ice only in the outer nebula, beyond about four AU (one AU, or "astronomical unit," equals the distance from the Sun to the Earth), a position sometimes called the "snow line." This boundary effectively divides the present solar system in two, with an inner region of relatively dry, rocky planets and an outer region of planets and moons made of both rock and ice, as well as icy comets.

The rocky cores of the outer planets are believed to have had similar origins to the rocky terrestrial planets; however, the outer planets also accreted vast amounts of ices from the billions of cometlike planetesimals that populated the outer solar system beyond the snow line. The two largest bodies, Jupiter and Saturn, also apparently accumulated large quantities of gas, mostly hydrogen and helium, directly from the solar nebula, giving them compositions that more closely approximate that of the Sun.

CLEANING UP

The assembly of planetesimals into planets basically swept the solar system clean of kilometer-sized objects. All that now remains of that once-vast population of planetesimals are a few thousand asteroids, prevented by Jupiter's massive gravity field from accreting into their own planet. But what of the nebular gas and small dust particles that were not included in planetesimals?

Hardly a trace remains now. Each cubic centimeter of space in the vicinity of the Earth's orbit contains only about ten atoms, many trillions of times less dense than in the original nebula. The energetic outbursts exhibited by young stars might provide an explanation for the missing matter. These protostars apparently remove their gaseous envelopes by violently blowing matter away at high velocities. Objects of meter size or larger would have resisted removal by such winds, but gas and tiny bits of dust would have been swept away. The result was a system swept clean of gas and dust—a gleaming new star, adorned only by a few revolving bright beads, the planets and their moons. Planetary accretion screeched to a halt, robbed of the necessary raw materials. This spring cleaning is thought to have happened quickly, in less than a million years.

HARMONY REVISITED

As celebrated in geometric figure and song by Johannes Kepler, the planets possess some regularity in their orbital spacings. In more recent times, Kepler's zany explanations for this relationship have been ignored, and the regularity is referred to as the Titius-Bode rule after two eighteenth-century scientists, one of whom expressed it mathematically and the other who popularized it. Broadly speaking, the rule states that each planet is twice as far from the Sun as its nearest interior neighbor. The relationship only holds approximately for the planets, although geometric spacings in the satellite systems of Jupiter, Saturn, and Uranus also seem to follow its formulation. It is now thought to result from gravitational and tidal evolution following the formation of the planets and satellites.

The harmony in the motions of planets sought by Kepler is apparently not an artifact of their origin, that is, not a consequence of forces and events in the solar nebula. The only vestige of that ancient nebula in planetary motions is the confinement of the planets to the ecliptic plane, a clue recognized by Newton. However, it was nearly a century before a perceptive Laplace used this clue to good advantage in devising his nebular hypothesis, and almost as long again before this idea was confirmed by astronomical observation.

The origin of the Earth must be considered in the context of the origin of the entire solar system. Ours is a world, like its neighbors, assembled piecemeal from planetesimals, huge boulders which in turn were molded from tiny grains of nebular condensate and interplanetary dust. The birth of planets happened within a dense fogbank of gas and dust, lit spasmodically by a violent young Sun. The real harmony of the planets lies not in chords played out by their Keplerian motions, but in the overtones and dissonances of their shared creation within the solar nebula.

Some Suggested Readings

Allegre, Claude (1992). *From Stone to Star: A View of Modern Geology.* Harvard University Press, Cambridge, MA. *Chapter 4 of this superb little book describes the condensation sequence, and Chapter 6 outlines the accretion of the planets in easily understood terms.*

Boss, A. P. (1990). 3D solar nebula models: Implications for Earth origin. In *Origin of the Earth,* edited by H. E. Newsom and J. H. Jones, Oxford University Press, New York, pp. 3–15. *This paper provides a more rigorous view of the dynamic and thermal evolution of the nebula; not for the casual reader.*

Gingerich, Owen (1993). *The Eye of Heaven: Ptolemy, Copernicus, Kepler.* American Institute of Physics, New York. *A masterful account of Kepler's attempt to find harmony in the motions of the planets.*

Podosek, F. A., and Cassen, Patrick (1994). Theoretical, observational, and isotopic estimates of the lifetime of the solar nebula. *Meteoritics,* vol. 29, pp. 6–25. *This excellent scientific paper summarizes the constraints on nebula formation from theory and astronomical observations, as well as analyses of meteorites.*

Taylor, S. R. (1992). *Solar System Evolution: A New Perspective.* Cambridge University Press, New York. *Chapter 2 of this thoughtful book provides a comprehensive summary of the origin and evolution of the solar nebula.*

Wood, J. A., and Morfill, G. E. (1985). A review of solar nebula models. In *Meteorites and the Early Solar System,* edited by J. F. Kerridge and M. S. Matthews, University of Arizona Press, Tucson, pp. 329–47. *An interesting contribution describing the changes that have occurred in nebula models during the last few decades and elaborating on the presently favored concept; not for the casual reader.*

Simmer Until Done

Differentiation of the Earth's Interior

THE IMPORTANCE OF HEFT

Deep inside the Earth there is a heart of metal, an iron sphere as large as Mars. This *core* is sequestered within a *mantle* of silicates and oxides sometimes squeezed by high pressure into structures so dense that they cannot exist at all on the planet's surface or within its outermost layer, the *crust*. It is not intuitively obvious how the world came to be this way. It seems unreasonable to suppose that the embryonic Earth began as a huge iron nugget, onto which flocked a mantle of silicates, because the iron did not condense first. Therefore, we must assume that the core, mantle, and crust somehow separated from each other during or after planetary accretion. Geologists use the term *differentiation* to describe the collective processes by which these distinctive regions originated.

A favorite dessert at my house is a sort of cherry cobbler. The recipe calls for distributing a layer of cake batter on the bottom of the pan, supplanted by a layer of cherries. As this mixture cooks, the cake rises and the cherries filter down through the cake to form a layer on the bottom. The inversion of the original layering results from the fact that cake batter becomes less dense as it transforms into fluffy cake, so that the heavier cherries sink, producing a gravitationally stable configuration.

Something similar to this process is thought to account for the Earth's differentiation, a process dependent on *density* (defined as mass per unit volume). Dense blobs of molten metal, originally distributed throughout the mantle, sank through the surrounding silicate matter to form a core in the early Earth. Conversely, partial melting of the silicate portion of the mantle yielded magmas that were less dense than the enclosing solids, and these magmas ascended toward the surface to form a crust. Although much of this activity happened very early in the planet's history, the interior still broils and matter continues to sink or rise, according to its density. As we will see, the ongoing evolution of our planet, indeed its geological (and probably biological) vitality, depends on the continued operation of its internal oven.

FLAME BROILED

In *Journey to the Center of the Earth,* published in 1864, novelist Jules Verne described our world as one riddled with holes, a view shared by many scientists of his day. These were not just caverns in the crust, but wormholes penetrating down to the very center of the planet. Some cavities were thought to be empty (that is how Verne's intrepid explorers could use them as passageways to the core), and others were thought to contain molten magmas derived from deeper levels. In Verne's imagination, these twisted hollows coalesced eventually at the Earth's center to form a gigantic void.

For centuries miners had recognized that underground mines were warm, and the deeper one went, the hotter it became. This observation led to the seemingly logical conclusion that the Earth's central void must be a simmering cauldron, perhaps filled with fire that caused the wall-rocks to melt (presumably this was the source of magma for volcanic eruptions). This idea was bolstered by theories suggesting the Earth was actually an aborted star, now cooled to the point where its skin had lithified. Even James Hutton, champion of the idea that igneous rocks were congealed magma, favored the hypothesis of a central fire to explain their occurrence. This notion persisted until the end of the eighteenth century.

But the concept of a central fire was eventually to be extinguished and the void in which the flames raged was to be filled with another kind of substance. In 1798 Lord Cavendish, using the balance period of a pendulum, calculated the mass of the Earth which, divided by its volume, yielded its *mean density*. His estimated value, 5.45 grams per cubic centimeter, is nearly identical to the presently accepted value of 5.25. This density estimate was surprisingly high. To provide some perspective, the familiar and seemingly hefty rocks in the Earth's crust have densities only about half that amount. The discrepancy can be accounted for by having very massive matter in the planet's interior, a requirement clearly inconsistent with an inflamed void. All manner of suggestions for the nature of the dense core material ensued, including the radical idea that our planet had a heart of precious metals like gold.

The existence of a dense core was also supported by precise measurements of the shape of the Earth. As a rotating ball turns on its axis, centrifugal force causes its equator to bulge. The Earth does this too, adopting a slightly flattened shape whose equatorial diameter is 1/300th longer than its polar dimension. The magnitude of the effect is most pronounced if mass is uniformly distributed throughout the interior and less if a greater amount of mass is located at the center, but it is measureable in either case. In the early nineteenth century, measurement of this property, called the *moment of inertia,* for the Earth confirmed that it must have a massive, central core.

GOLDSCHMIDT'S RECIPE

During World War I, Norway was largely isolated from the rest of Europe, so that it lost access to raw materials vital to its industries. At the behest of the Norwegian government, Victor Goldschmidt, a young geology professor at the University of Oslo, established a laboratory to study how these missing raw materials might be replaced by other natural materials native to the country. This led him to investigate the chemistry of economic minerals and, eventually, the more general relationships between a substance's chemistry and its crystal structure, work for which he is now known as the father of geochemistry.

In 1922 Goldschmidt published a short but important paper (in German) entitled "The Differentiation of the Earth," in which he stated:

> It is conceivable that the original state of the Earth was a homogeneous or nearly homogeneous mixture of the chemical elements and their compounds. Today, however, the Earth is far removed from a homogeneous state. The material distribution within the Earth has by no means reached a final equilibrium state; we observe instead an active redistribution of matter and energy. The processes which have resulted in the inhomogeneity of our

This Norwegian postage stamp, issued in 1974, honors geochemist Victor Goldschmidt.

> planet and still contribute to the migration of material I would summarize in the expression "the differentiation of the Earth."
>
> —V. M. Goldschmidt, *"Der Stoffwechsel der Erde"*

In proposing this new term, Goldschmidt focused attention on the processes by which the core, mantle, and crust formed. He suggested that the Earth had originally been a molten ball, consisting of silicate, metal, and sulfide. Gravity caused the liquid metal, being more dense, to sink and form a central core. Molten sulfide, less dense than metal but denser than silicate liquid, likewise sank to form a shell surrounding the metallic core, leaving the melted silicate behind to comprise the mantle. The segregation of metal and sulfide from molten silicate he termed the primary geochemical differentiation. Then, as the global ocean of silicate liquid cooled and solidified, the denser silicate minerals, mainly olivine and pyroxene, that formed from it sank and the less dense minerals, feldspars and quartz, floated to the surface. This secondary differentiation formed the mantle, rich in silicates of magnesium, and the overlying crust.

Experimental data on which to base this hypothesis were meager, but Goldschmidt suggested that meteorites were natural experiments with molten metal, sulfide, and silicate, now in a frozen state. From the chemical analysis of these individual phases in meteorites, Goldschmidt was able to understand how different elements partitioned among these substances, and he coined terms to describe elements with pronounced affinities for metal (*siderophile*), sulfide (*chalcophile*), and silicate (*lithophile*).

More important, however, understanding an element's geochemical affinity explained how it might have been distributed between the core, mantle, and crust during the Earth's differentiation. For example, nickel, a siderophile element, was concentrated in the metallic core, accounting for its paucity in crustal rocks. The use of meteorites as proxies for the Earth's composition was a magnificent conceptual leap, one that even today forms the basis of modern ideas about the composition of the planet's interior.

EARTH TONES AND SHADOW ZONES

Goldschmidt's need for a recipe for planetary differentiation was guided by the arguments from mean density and moment of inertia that the Earth's interior must contain some dense material, but also by more tangible proof that a core existed. This proof came not from direct sampling of the interior, but from the indirect observations of *seismology* (the study of vibrations caused by earthquakes). We might liken this approach to determining when a cake is done by the change in sound produced by knocking on the side of the pan.

A journey to the center of the Earth requires traversing approximately 6,370 kilometers. Virtually all of this distance is unknown territory. The deepest wells ever drilled by man extend no more than ten kilometers down, and chunks of mantle rock rafted upward by volcanic eruptions generally come from depths of less than a few hundred kilometers. The mean density and moment of inertia of the Earth tell us that its mass is not distributed uniformly, but the exact structure of the interior and the composition of its parts cannot be determined from these measurements. So how do we know what's down there?

The key to answering this question is wiggles and jiggles—precisely metered seismic waves. In 1875, John Milne, an English mining engineer, visited Japan as a member of a team bringing Western technology and education to that isolated country. Japan's numerous earthquakes caught Milne's interest and, within a few short years, he became a pioneer in seismology. Largely through his efforts the seismograph, an instrument for recording seismic vibrations, was transformed from a curiosity into a precise analytical tool. After gauging their energy, Milne made the astonishing prediction that Japanese earthquake vibrations might be detected anywhere around the globe.

The rumblings that jostle seismographs actually emanate from deep underground faults. Seismic waves that travel upward create havoc on the surface; those that move downward through the planet may eventually emerge elsewhere, to provide information on the interior structures that they encountered. The speed of a seismic wave depends on the density and elasticity of the materials through which it travels. When the wave suddenly comes upon material with new properties, it may either be reflected back toward the surface or be bent (refracted) as it enters the new medium.

It is this refraction of seismic waves that demonstrates that the Earth has a core. Seismic waves reaching a depth of 2,900 kilometers suddenly come upon a new substance that is much denser than the overlying rock. At this boundary, called the Gutenberg discontinuity after its discoverer, some of the energy is reflected back to the surface, but some continues downward and, as it does so, its trajectory is strongly bent. The result is a loss of signal on the other side of the Earth, a region 10,000 to 16,000 kilometers away from the earthquake's location, a distance equal to about 41 degrees of the planet's circumference. It is as if seismic signals on that portion of the globe were in the shadow of some large obstacle. This seismic *shadow zone* is the ghostly image of the Earth's core. At larger arc distances the seismic waves are again received, but their travel times are longer than expected, indicating that they were slowed during their passage through the planet's center. The actual discovery of the core in 1906 was based on detailed analysis of the travel times of seismic waves, work done by Richard Oldham of the Geological Survey of India.

The composition of the core is now understood to be mostly iron metal, as determined from several kinds of reasoning. The core comprises approximately one third of the mass of the Earth, and iron is the only dense element abundant enough in solar system matter (as estimated from the composition of chondrites) to compose it. The mantle and crust actually contain only a small fraction of the iron predicted to occur in the Earth, and a metallic iron core is required to reconcile the planet's mean density with its inferred bulk composition. Moreover, iron meteorites, the twisted chunks of metal on display in many museums, demonstrate that this element formed cores in other differentiated bodies.

The experimentally determined velocity of seismic waves traveling through iron metal also provides an acceptable match to seismic velocities measured for the core, at least when the iron is held at high pressure. This measurement was a triumph of experimental ingenuity. Francis Birch, a professor of geophysics at Harvard, spent his career painstakingly devising ways to measure seismic velocities in various materials at ele-

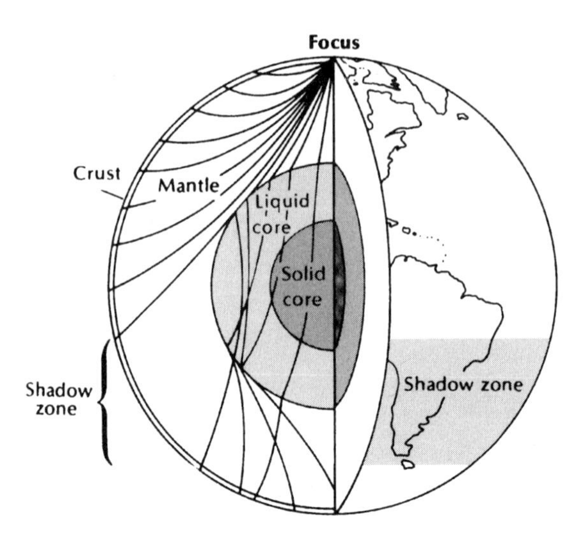

An earthquake originating at the North Pole would create a shadow zone, as illustrated here. Seismic waves travel downward until they intersect the core. At that point the waves slow down abruptly and bend toward the Earth's center, so that seismic signals do not appear on the planet's surface opposite this region. Beyond 143 degrees of arc from the location of the earthquake, seismic waves are again received.

vated pressures and temperatures. Laboratory experiments under such conditions are difficult and dangerous, but Birch's meticulous research finally bridged the gap between seismology and mineralogy. His work also demonstrated that an alloy of iron with nickel provides a better match with core velocities than does iron alone. The inference that the Earth's core is iron-nickel alloy is supported by the observation that iron meteorites also contain small quantities of nickel.

But more details about the core than just its composition can be extracted from seismicity. There are actually two kinds of seismic vibrations, one called *P* (or "primary," because it travels fast and arrives first) waves and another called *S* (or "secondary," because of its later arrival) waves by Oldham. The S waves, which are unable to pass through a liquid, are observed to disappear at the top of the core. (The shearing motion of S waves cannot be transmitted through liquids because their loosely bonded molecules slip past each other too easily.) Then, at a depth of about 5,000 kilometers, the P-wave velocity increases abruptly and there is a hint of the reappearance of an S wave. From such observations, Danish geophysicist Inge Lehman hypothesized in 1936 that the core must be stratified, with the outer portion liquid and the center solid. The existence of liquid metal at these depths requires some alloying light element (possibly sulfur or oxygen) to serve as an antifreeze, because pure iron-nickel should be solid under the conditions of the outer core.

The mantle's structure and composition, too, can be studied with seismic waves. Its upper boundary, called the Mohorovicic discontinuity (the abbreviated version "Moho" allows seismologists to discuss their work at cocktail parties), separates the crust from denser mantle rocks below. The identities of the minerals comprising mantle rocks can be inferred from the speeds with which seismic waves travel through them. The silicate minerals, mostly olivine, pyroxene, and garnet, that comprise the upper mantle undergo a series of structural changes with depth, as the pressure steadily increases and various minerals reach their limits of stability. At 400 kilometers and then again at 670 kilometers, new denser crystalline substances appear, resulting in sudden jumps in seismic wave velocity at these levels. The upper discontinuity represents the transformation of olivine into the spinel structure which, in turn, inverts at even greater depth to perovskite, a dense mineral stable only under extreme pressure. Although no mantle perovskite will ever be found in crustal rocks or grace a single mineral collection, its abundance in the lower mantle qualifies it as the Earth's most common mineral. Below about 670 kilometers depth, no further sudden changes are noted, though the lower mantle continues to undergo compaction and seismic wave velocities slowly increase.

The horizon separating the core and mantle appears to be a particularly complex region. The bottom several hundred kilometers of the mantle are given the name *D*" (pronounced "*D* double prime"), a vestige of an earlier seismological nomenclature that labeled all the Earth's recognized layers sequentially from *A* (the crust) to *G* (the inner core). In this system the lower mantle (below 670 kilometers depth) was *D;* sometime later, the lower mantle was subdivided into regions recognized to have distinctive properties, and the portion just above the core was designated *D*". This alphabetized nomenclature has now been abandoned, except for the peculiar *D*". Anyway, this region is characterized by large fluctuations in the speeds of seismic waves, which are interpreted to indicate radical differences in temperature caused by interactions between the lower mantle and outer core.

The crust is only a thin veneer, nowhere thicker than a hundred kilometers and commonly closer to 10 or 20 kilometers thick. Crustal rocks are significantly less dense than mantle rocks. The crust's characteristic seismic velocity is appropriate for rocks we know well from

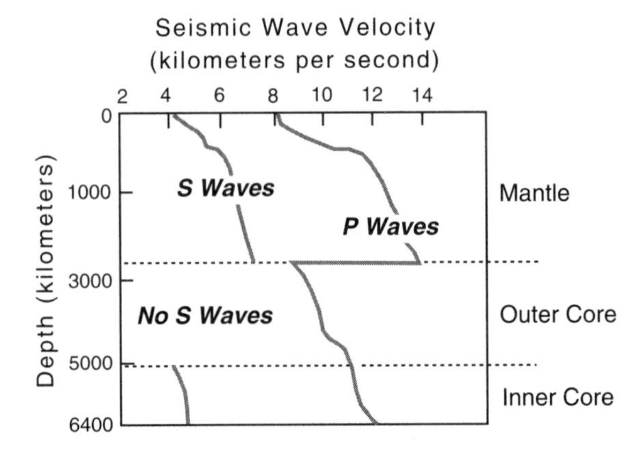

An internal profile of the Earth, showing how P and S seismic wave velocities change with depth. The greatest discontinuity in P wave velocity is at the core-mantle boundary, and S waves are not transmitted through the liquid outer core. At this scale, the crust is so thin that the Moho (the discontinuity at the crust-mantle boundary) is not resolved.

studying the surface geology—granite, gabbro, and the sediments derived from them.

Seismology has revealed many details about the differentiation of the Earth into crust, mantle, and core. The crust on which we live is only a thin rind, very different in composition from the mantle and unrepresentative of the planet as a whole. The mantle comprises the bulk of the Earth. It consists mostly of a succession of magnesium-rich silicates and oxides, the exact mineralogy depending on depth. The core, composed of metallic iron-nickel alloy, is part liquid (the outer core) and part solid (the inner core). The core-mantle boundary appears to be a dynamic region where these two very different kinds of matter interact.

HALF-BAKED AT THE START

From knowledge that the Earth formed by accreting small planetesimals, we might logically infer, as did Goldschmidt, that our planet was homogeneous throughout at the time of its birth. That would require a great many planetesimals, themselves homogeneous, all with basically the same chemical and mineralogical composition. Goldschmidt suggested that these planetary building blocks were chondritic meteorites.

That was a very plausible guess. The chemical compositions of chondrites are so close to that of the Sun (and therefore the average solar system composition) that it is reasonable to suppose that planetesimals of chondritic composition were widespread in the early nebula. We also have now confirmed, although it was not known in Goldschmidt's time, that chondrites actually formed within the solar nebula and that their original compositions have been preserved intact from that early epoch. Goldschmidt recognized that chondrites are, for the most part, intimate mixtures of metal, sulfide, and silicates, so they themselves have the right basic ingredients to make our planet. In fact, the bulk chemical composition of the Earth appears to be basically similar to that of chondrites.

This last assertion requires some elaboration. Because the Earth is differentiated, there is no place on it or within it where we can obtain a rock that has the composition of the whole planet. How, then, can we say that the bulk composition of the whole Earth is chondritic? This conclusion is based on the observation that *ratios* of the concentrations of certain elements in terrestrial rocks are identical to those in chondritic meteorites. These particular elements all behaved in precisely the same manner during differentiation and during later geologic events. As an example, let's consider the geochemically similar elements uranium and lanthanum. If uranium was concentrated or depleted by some process, lanthanum was as well, in the same manner and to the same degree. Thus the ratio of uranium to lanthanum remains fixed at its starting value, that is, the value of the whole Earth, although the absolute concentrations of both elements may vary from rock to rock. Sure enough, thousands of terrestrial rock analyses demonstrate that their measured uranium-to-lanthanum ratios are constant, and this ratio is the same as for chondrites.

The volatile element potassium also behaves in a geochemically similar way to uranium and lanthanum, both of which are refractory. However, the abundance ratio of potassium to either uranium or lanthanum is significantly lower in terrestrial rocks than in chondrites. Other volatile elements are similarly depleted in the bulk Earth. So, we can infer that the Earth basically has a chondritic bulk composition for refractory elements but it is missing some of the volatile elements, either because they never fully condensed into the planetesimals that formed the Earth or because they were somehow lost during planet formation.

One popular view of how differentiation occurred, favored until just a few years ago, has been variously termed the "iron catastrophe" or the "big burp." In this model, the Earth accreted from cold, chondritic material. Subsequent heating of this homogeneous globe resulted from accretion, compaction, and especially the slow decay of radioactive isotopes of uranium, thorium, and potassium in the planet's interior. Such long-lived isotopes release heat very slowly, but they can be a very effective source of heat over billions of years (these isotopes continue to decay, even today, inside the planet). If all the heat from decay of the radioactive isotopes in a thumbnail-sized lump of granite were retained, it would take millions of years to brew a cup of coffee, but over several billions of years the rock itself would finally melt. So after a long time, perhaps a billion years, the Earth's internal temperature would have reached the melting point of iron metal, allowing the dispersed metallic grains to have melted, pooled, and finally segregated to form the core.

The plausibility of this model has been questioned, based on some experiments that suggest that segregating molten metal from silicates may have been difficult. Liquid in contact with solid can either form beads or a thin film, as raindrops do when falling on the hood of a car (depending on whether or not the car's surface is waxed). To form a core from a homogeneous chondritic mantle, molten metal droplets had to migrate through the silicate mesh and collect in pools large enough to sink. The ability of liquid metal to segregate in this way depends on whether or not it "wets" the solid surface. The problem remains unresolved, but in any case, a revised model for planetary differentiation may render the question of how efficiently small beads of metal migrate irrelevant.

The new model is based on a reevaluation of the nature of the planetesimals from which the Earth was assembled. Because the Earth's bulk composition is the same as chondrites, we have assumed (as did Goldschmidt) that the bodies from which it accreted were chondritic. Paradoxically, that assumption is both right and wrong. To make sense of this statement, we need to understand where chondritic meteorites formed.

Located between the orbits of Mars and Jupiter is the asteroid belt, containing thousands of relatively small chunks of matter, mostly ranging in size from a few kilometers to a few hundred kilometers in diameter. These bodies are surviving planetesimals. Asteroids are mere pinpricks of light (the name literally means "starlike"), but unlike stars they reflect the light of the Sun rather than shining with their own light. They do not, however, reflect all the incident sunlight uniformly. Some wavelengths of light (colors) may be absorbed by minerals on an asteroid's surface, which can be discerned from its *spectrum* (the measured amounts of light reflected at different wavelengths). Distinct classes of asteroids are distinguished based on their spectra, and the distributions within the asteroid belt of hundreds of classified objects have now been mapped. The orbits of each asteroid class seem to be roughly clumped at similar distances from the Sun.

Most meteorites are thought to be pieces of asteroids, and possible links between various meteorite types and specific asteroid classes have been proposed. By comparing the spectra of meteorites obtained in the laboratory with telescopic spectra of asteroids, it is possible to recognize objects with common mineralogy. From the meteorites, we can then infer the thermal histories of various asteroid classes.

The inner asteroid belt (containing bodies located closer to the Sun than 2.7 AU) mostly consists of asteroids containing minerals (olivine, pyroxene, and metal) that form at high temperatures. Variations in the proportions of these minerals are indicated by differences in the spectra observed as the asteroids rotate, implying that they are actually heterogeneous bodies that have been broken apart by impacts, exposing different compositions in their interiors. The most straightforward explanation for their mineralogic heterogeneity is that they have been melted at high temperatures, at least 1,100 degrees Celsius. Segregation of core material or magma from unmelted crystals then resulted in different parts of the asteroids having distinct compositions. Some bodies in this region appear to be just hunks of metal, cores that have completely been stripped of their rocky mantles by repeated impacts. Asteroids in the inner belt once may have been chondritic, but most of them are now differentiated. These bodies are apparently the sources of igneous meteorites (called *achondrites*) and iron meteorites. The ages of achondrites and iron meteorites indicate that asteroid differentiation happened very quickly after their formation.

Near the middle of the asteroid belt are bodies with spectra suggesting that they have been heated and recrystallized at modest temperatures (geologists call such a process *metamorphism*), but not melted. The spectra of some asteroids in this region also indicate the presence of clay minerals. Clays are hydrous phases that form when olivine and pyroxene are altered in the presence of water. Water bound into the structure of clay vibrates when it absorbs light of a particular wavelength, so its spectrum is diagnostic. If these bodies had originally accreted ices along with rock, and the ice later melted, the formation of clays might be explained. Melting ice requires only a modest temperature increase, something above 0 degrees Celsius. The inferred mineralogy of asteroids in the middle of the belt is similar to chondritic meteorites, and this is probably the region from which chondrites are derived.

Even farther out, past 3.4 AU, are an assortment of very dark asteroids. From their spectra we infer that they contain rock and ice, but no hydrous clay minerals. So at the far reaches of the asteroid belt, ice-bearing planetesimals were apparently never heated even to the melting temperature of ice. These "iceteroids" are apparently not represented in meteorite collections, although some interplanetary dust particles may be samples of them.

To summarize, the picture of the asteroid belt that emerges from spectral studies is as follows: Objects nearer the Sun were heated to temperatures high enough to melt silicate rock and metal. Objects farther out experienced lower temperatures, hot enough to cause metamorphism or to melt ice for the formation of clay minerals. Asteroids at even greater distances remained cold enough to preserve frozen ice.

One possible explanation for this pattern is heating by the rapid decay of radioactive aluminum-26. The former existence of this isotope has been demonstrated in meteorites and, if present in sufficient quantity, it would have been a potent heat source. The rate of accretion would have been fastest where there was more matter to accrete, and nebula models indicate that the density of gas and dust was greater toward the center. The earlier a planetesimal formed, the greater the proportion of live aluminum-26 it would have contained, and so the hotter it would eventually become. Planetesimals nearer the Sun should have accreted earlier and melted, whereas those at larger distances should have formed later and been heated only slightly or not at all. Electromagnetic induction has also been proposed as an asteroid heat source. This mechanism requires a ferocious solar wind with a strong magnetic field, such as is envisioned for infant T Tauri stars. The magnetism is converted into electric currents inside planets and asteroids, and resistance to the flow of these currents produces heat. However, the recognition that most of the solar wind in such stars is expelled from the poles, rather than out through the nebular disk, makes this idea less likely.

Whatever heat source one favors, the relevant point gleaned from asteroid heating is this: most of the planetesimals from which the Earth accreted must have formed even closer to the Sun than the inner asteroid belt, so they would certainly have been melted and differentiated. In effect, they would have been mini-Earths, with cores, mantles, and crusts. The effect of accreting already differentiated planetesimals would

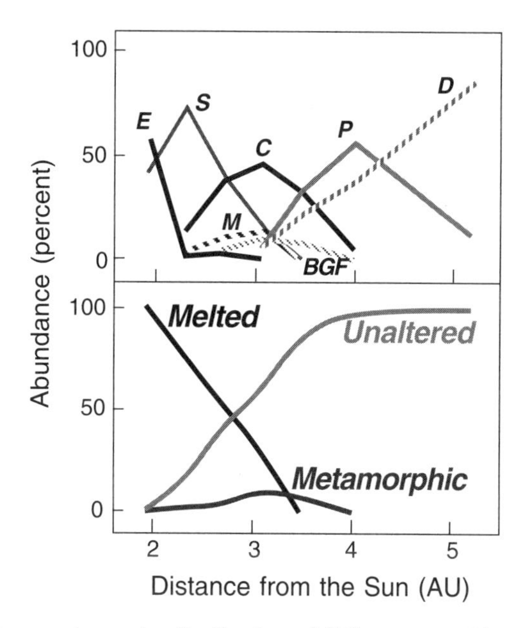

The upper diagram shows the distribution of different asteroid spectral classes (designated by letters) with increasing distance from the Sun (expressed in AU; one AU equals the distance from the Sun to the Earth). The bottom diagram provides an interpretation of the thermal histories experienced by these asteroid classes, formulated from studies of meteorites that are spectrally similar to asteroids and are thought to be samples of them. (Modified from J. F. Bell, D. R. Davis, W. K. Hartmann, and M. J. Gaffey [1989], *Asteroids II,* University of Arizona Press)

have been to accelerate the process of core formation in the growing Earth, because metal already existed as large, dense cores rather than as tiny dispersed grains. If planetary accretion was fast enough, the planetesimals might have still been molten or partially molten at the time they were accreted. Alternatively, the violence of planetesimal accretion would probably cause some remelting. In either case, heavy blobs of metallic liquid would have rapidly sunk through the Earth's mantle to form the core.

Even though the accreting planetesimals were differentiated, they still retained chondritic *bulk* compositions, explaining why the Earth does as well. The low concentrations of volatiles in achondrites, relative to chondrites, suggests that the accreting objects were already depleted in volatiles. Therefore, our planet's volatile element abundances may be an inheritance from the differentiated planetesimals that formed it, rather than a loss that happened after its assembly.

PLUMES AS MIXERS

As recognized by Goldschmidt, the Earth's differentiation is an ongoing process, continually driven by heat from its internal oven. To a first approximation, the most important geophysical process that the Earth experiences is loss of heat. The yearly loss of internal heat energy is on the order of 10^{21} joules, roughly the power required to run 500 light bulbs continuously for every person on the planet. This prodigious amount of energy dwarfs those of other natural processes such as earthquakes and volcanic eruptions. As we have seen, continued heating inside the planet results from radioactive decay, but also from another source whose importance is just being fully appreciated. The liquid outer core is continually crystallizing, so that over time it shrinks as the solid inner core grows. Crystallization at the interface between the inner and outer cores releases a surprising amount of heat, which stirs the molten liquid metal of the outer core, in turn creating the Earth's magnetic field. The swing of a compass needle thus depends on heat generated at the planet's remote center.

The heat carried upward through the outer core is ultimately transmitted to the mantle interface, the D" layer. The core is much hotter than the overlying mantle, on average perhaps by 500 to 1,500 degrees, and this temperature difference is achieved across the width of D". Raymond Jeanloz, a geophysicist at the University of California at Berkeley, has suggested that perovskite, the primary constituent of the lower mantle, reacts vigorously with liquid metal at the high pressures and temperatures of the core-mantle boundary. If that is true, D" may be the most chemically active region of the planet. It is not surprising, then, that this layer shows oscillations in seismic velocity, reflecting temperature variations, over different parts of the core. What is unexpected is that these temperature variations correlate with the geology of the Earth's surface.

The core-mantle boundary is physically connected to events on the planet's surface by plumes of plastically deforming rock. Just above upwelling hot spots in the outer core, the D" mantle rock bulges like the cap of a mushroom. Eventually it becomes hot enough to rise buoyantly toward the surface. The rock is still solid, but its temperature is so high

These computer simulations illustrate conditions inside the Earth's mantle. In the image on the left, hot mantle plumes rise from the core-mantle boundary toward the surface. In the image on the right, cold mantle rock sinks from the surface and accumulates at the 670-kilometer discontinuity, occasionally penetrating farther down to the core. (Computer simulations by P. Tackley, California Institute of Technology)

that it can creep, like asphalt on a hot summer day. This plume remains connected to the D" boundary by a narrow pipe, which continues to feed hot rock to the ascending cap. As the plume approaches higher levels (and thus experiences lower pressures) the rocks begin to melt, and vast outpourings of basaltic lava flood the surface.

Above places where the core is cooler, slabs of crustal rock sink deep into the mantle. Some of these slabs halt their descent at the 670-kilometer discontinuity, whereas others appear to continue all the way down into D" and come to rest on the core-mantle boundary. Once there, the slabs spread out laterally and their weight distorts the boundary itself, like an upside-down mountain range.

Mantle plumes, which originate at the core and extend nearly all the way up to the surface, provide an effective way to move material and heat energy from the Earth's deep interior. Evidence for the descent of crustal slabs into the mantle indicates that materials go both ways. Over time, this global stirring must recycle rocks from the crust back into the mantle, in effect reversing the differentiation process. Iron-nickel metal, however, once sequestered in the core, remains there forever.

ISOTOPES, SHAKEN NOT STIRRED

Running along the floor of the Atlantic Ocean, like a gigantic spine, is a long mountain range. This spine, the Mid-Atlantic Ridge, is the site of effusive eruptions of lava. The basalts are generated by partial melting of the Earth's upper mantle, defined as the mantle region above the 670-kilometer discontinuity. All along the volcanic ridge, huge plates of crustal and upper mantle rock are pulled apart, so that magma from below wells up in the newly created fractures and congeals to form new seafloor. The Mid-Atlantic Ridge is only one of many such volcanic chains that encircle the Earth like seams on a baseball.

The ocean-floor rocks formed at mid-ocean ridges are, at least indirectly, samples of the mantle. They carry indelible fingerprints of their upper mantle source region, in the form of trace quantities of radiogenic isotopes (formed by the decay of radioactive parents). Radiogenic isotopes of strontium, neodymium, and lead are especially informative, and at least five distinct mantle sources have been identified based on diagnostic mixtures of these isotopes. Ocean-floor basalts (and thus their mantle source) have distinct strontium, neodymium, and lead isotopic compositions from the continental crust. However, adding the upper mantle and the crust together in the right proportions yields an intermediate isotopic composition that matches chondritic meteorites. This suggests that the crust has been extracted from an upper mantle that was originally chondritic, a geochemical demonstration of the planet's differentiation.

Hawaii and most other oceanic islands are huge basaltic volcanoes whose mantle source is clearly different from that of ocean-floor basalts. Their radiogenic isotopes indicate a mantle source region that is still chondritic; or in other words, it has not had crustal materials extracted from it. The basalts erupting from these volcanoes are thought to have been derived by melting of the lower mantle, rocks that originally resided at depths greater than 670 kilometers but now have been brought closer to the surface via plumes. Their isotopic compositions are complicated by mixing with contributions from ancient subducted slabs of ocean crust that they may have encountered during ascent, but the isotopic fingerprint of the lower mantle is clear.

These apparent differences in the composition of the upper and lower mantle imply that the 670-kilometer discontinuity may be an important boundary separating a highly differentiated upper mantle from a less differentiated lower mantle. (Actually, the lower mantle has had core

Basalts and Mantle Sources

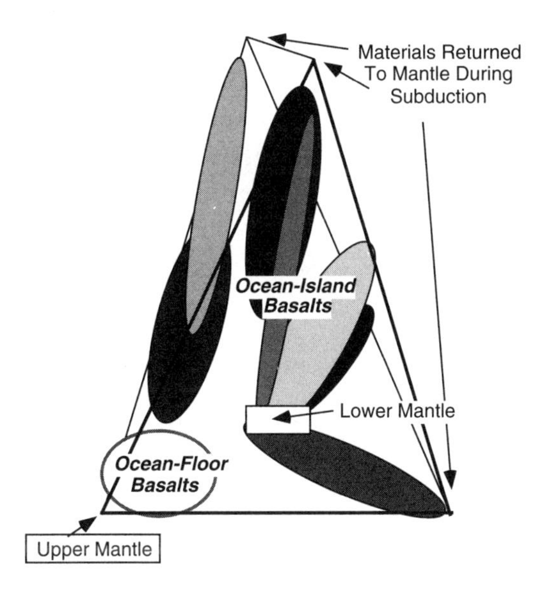

This diagram is a projection of a three-dimensional plot of the diagnostic ratios of strontium, neodymium, and lead isotopes measured in basalts from the ocean floor and from volcanic islands. Five distinct mantle sources are identified—four defining the corners and a fifth (the lower mantle) lying within the tetrahedron. Ocean-floor basalts, derived from melting the upper mantle, are distinct from ocean-island basalts, which form by melting the lower mantle. The latter are carried upward by plumes, where they often encounter and mix with subducted slabs of oceanic crust.

material removed from it, so in that sense it is differentiated too, but it apparently did not supply material for construction of the crust.) Seismic data further suggest that the lower mantle may be several percent denser than the upper mantle, partly as a result of slight differences in composition. The persistence of such chemical heterogeneity requires that the upper and lower mantles not mix very well. This isotopic picture seems contradictory to the view of mantle plumes emerging from seismic studies, and these two ways of looking at the mantle have not yet been fully reconciled.

RARE OR WELL DONE?

Differentiation is the most profound modification the Earth has undergone since its birth. The formation of the core, mantle, and crust represent irreversible changes in its very substance, wrought by heat and differences in density.

The great bulk of the work of differentiation is now done, but heating persists and so does internal change. Goldschmidt was right in arguing that matter and energy continue to be redistributed inside the planet, but he could not have dreamed that the scale of recycling would be so grand. Were it not for the great heat engine of the slowly crystallizing core and the slow decay of radioactive isotopes in the mantle, our world would be geologically dead, like its Moon. Quaking ground, volcanic mountains, the constantly shifting sculpture of the Earth's surface—all owe their existence to heat derived from the deep interior of the planet. And without continued release of heat, the fragile environments necessary for life on the surface would certainly collapse. The Earth is neither rare nor well done, but something in between, a concoction that thankfully still simmers on the stove.

Some Suggested Readings

Anderson, D. L. (1992). The Earth's interior. In *Understanding the Earth,* edited by G. Brown, C. Hawkesworth, and C. Wilson, Cambridge University Press, New York, pp. 44–66. *This is a particularly good example of the quantitative descriptions of the Earth's interior available in modern geology texts.*

Dziewonski, A. M., and Anderson, D. L. (1984) Seismic tomography of the Earth's interior. *Scientific American,* vol. 72, pp. 483–94. *A excellent overview of the breakthough technique that allows the three-dimensional structure of the mantle and core to be imaged.*

Hazen, R. M. (1993). *The New Alchemists.* Random House, New York. *A wonderful history of the technological development and scientific discoveries of high-pressure research applicable to understanding the Earth's interior. The book tends to focus on making diamonds, but also provides a rich mine of information on other high-pressure minerals.*

Jeanloz, R., and Lay, T. (1993). The core-mantle boundary. *Scientific American,* vol. 268, pp. 48–55. *A detailed description of possibly the most bizarre and chemically active place on Earth.*

Mason, Brian (1992). *Victor Moritz Goldschmidt: Father of Modern Geochemistry.* Geochemical Society Special Publication No. 4, San Antonio. *This interesting book describes the life and scientific contributions of the founder of geochemistry and his pioneering work in understanding the Earth's differentiation.*

Newson, H. E., and Jones, J. H., eds. (1990). *Origin of the Earth.* Oxford University Press, New York. *Several chapters in this excellent though technical book offer interesting perspectives on planetary differentiation. Especially relevant is a section by S. R. Taylor and M. D. Norman that describes the evidence for accreting already differentiated bodies, as well as one by D. J. Stevenson on the mechanisms of core separation.*

Wysession, Michael (1995). The inner workings of the Earth. *American Scientist,* vol. 83, pp. 134–47. *A wonderful and up-to-date overview of the differentiation of the Earth and geophysical constraints on the nature of its interior.*

A Pummeled Earth and a New Moon

Impact Disruption and Sometimes Reassembly

A HARD LANDING

As an Air Force pilot, my assignment was to fly the C-141 Starlifter, a four-engine jet transport plane, on routes all over the globe. Mostly we shuttled pallets of supplies and occasionally troops between military installations, but my squadron also had responsibility for transporting armored limousines for the White House and mail for various United States embassies, so we landed at civilian airports too. We used to joke among ourselves that our job was to fly the unnecessary and the unwanted to the ungrateful.

With its wheels and flaps extended, the Starlifter is a lumbering, 162-ton behemoth but, if piloted with sufficient skill, it is capable of landing with infinite grace. It was not uncommon for other crew members to test the prowess of a young pilot such as I by balancing a thin cigarette lighter on the console beside me just before touchdown. Without a word spoken, I realized that a gauntlet had been thrown down—a challenge to land the aircraft and bring it to a halt so smoothly and gently that the lighter would not tip over. This game obviously tested my ability to deftly control the aircraft and, perhaps not so obviously, to read and correct for the motion of the air mass within which the plane traveled. Thankfully there was never a need to worry about the movement of the runway itself. Nonetheless, it is a fact that the runway, situated as it is on the surface of the Earth, is hurtling through space at about 70,000 kilometers per hour. We Earth dwellers are blissfully unaware that our home planet is really a gigantic projectile. The only thing that makes a gentle, controlled collision between aircraft and runway so routine is that the plane and the Earth move through space together as one.

But what happens when two massive objects moving independently in space come together? We have already learned that dust and gas in the solar nebula clumped into small planetesimals which, in turn, gravitationally accreted to form the larger asteroids, moons, and planets. In such a hierarchical process, with little blocks coming together to form ever larger ones, it was inevitable that the last stages of planetary accretion must have involved the collisions of some very large bodies, sometimes with catastrophic consequences.

The Earth itself experienced many such cataclysms early in its history, and these events profoundly affected its evolution. However, billions of years' worth of geologic processes have all but erased the visible evidence of such terrible collisions. In order to reveal this awesome secret of our planet's past, we must turn for understanding to its neighbors—bodies whose more limited geologic processing permits evidence of their early histories to have been preserved. We will begin our foray by examining some rather small solar system objects, and then proceed to larger ones.

ANGELINE'S CRATER

Angeline Stickney was the maiden name of the wife of Asaph Hall, an American anstronomer who in 1878 discovered the two tiny moons of Mars. Hall christened his discoveries Deimos and Phobos, after the horses that drew the chariot of the god of war. Angeline was, at the very least, a very determined woman, certainly supportive of her husband's career and perhaps even something of a nag. "The chance of finding a satellite appeared to be very slight," wrote Hall, "so that I might have abandoned the search had it not been for the encouragement of my wife." Nearly a hundred years later, the *Mariner 9* spacecraft provided the first clear images of Hall's tiny moonlets. The most noticeable feature seen on either body is a prominent circular bowl 11 kilometers across, carved into the surface of Phobos, a potato-shaped moon that is itself only 27 kilometers in longest dimension. Belatedly but appropriately acknowledging Mrs. Hall's role in the discovery of this moon, the International Astronomical Union adopted "Stickney" as the name for this crater.

Radiating from Stickney crater in all directions are deep grooves. In mapping these furrows, planetary scientists Peter Thomas and Joe Ververka of Cornell University discovered that they converge on the other

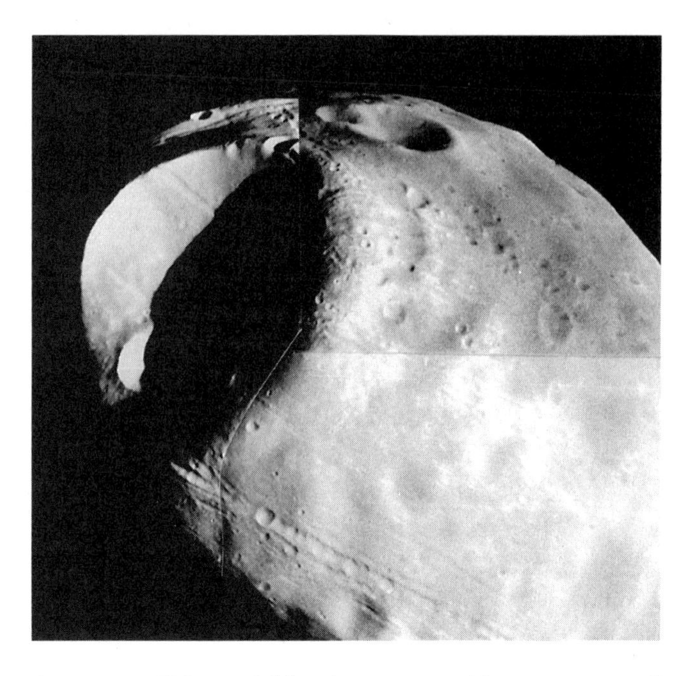

Phobos, a tiny moon of Mars, exhibits a huge crater with grooves extending radially in all directions. The impact that created this crater nearly tore Phobos apart. (Jet Propulsion Laboratory)

side of Phobos at a point directly opposite the crater, a location called the *antipode*. They suggested that the impact that formed Stickney crater also produced cracks in the satellite, and may even have generated enough heat to release steam from the interior. This hot vapor, derived from dehydrating water-bearing minerals, would presumably have enlarged the grooves as it escaped through them to space. More recent computer modeling of the propagation of interior cracks resulting from the Stickney impact supports the idea that this impact was nearly catastrophic. The models suggest that the radial furrows emanating from Stickney crater are not merely surface features, but must be cracks that penetrate through the body itself. This impact, seemingly modest compared to those that produced larger craters on many other worlds, almost demolished tiny Phobos.

ALL IN THE FAMILY

The realization that craters like Stickney were formed by impacts has been slow in coming. Generations of astronomers and geologists be-

This photograph, taken by G. K. Gilbert in 1891, shows one of his early impact experiments with clay impactors and targets. (U.S. Geological Survey)

lieved that craters on the Moon were volcanic features, and the issue was settled only in this century. Leading the vangard was G. K. Gilbert, chief geologist of the U.S. Geological Survey. In 1891, Gilbert conducted the first impact cratering experiments, using balls of clay to impact a clay target. He was particularly interested in the shapes of craters and was able to demonstrate that impactor velocity controlled the morphology of the excavated hole.

Curiously Gilbert, champion of the impact hypothesis for lunar craters, was unwilling to accept the idea that the very same process could have formed a crater on the Earth. Daniel Barringer, owner of the Arizona desert property on which Meteor Crater is situated, argued that this huge bowl was the scar of an impacting meteor. Barringer, a mining engineer, was motivated by a desire to find and recover the huge mass of iron-nickel metal that he thought was lodged in the bottom of the crater, and financial backing depended on convincing his bankers that a buried meteorite was really there. Gilbert visited the site once, pronounced it the result of a steam explosion, and thereafter ignored Barringer's persistent efforts to explain its origin through impact. He dismissed the tons of iron meteorite fragments scattered around the crater as coincidence. Gilbert's influence among the community of geologists was so great that decades went by before Barringer was finally vindicated.

The mechanics of crater formation are now reasonably well understood, thanks to modern laboratory experiments using large guns that fire projectiles into stationary targets and to the study of artificial craters produced during nuclear bomb tests. However, punching holes in soil and rock, however violently it may be done, is not the same thing as the actual breaking apart of bodies during collisions, and our understanding of the physics of disruption is on less firm ground. To model the collisional fragmentation of even small solar system objects, say the size of Phobos, we must extrapolate from more modest laboratory studies with targets the size of a marble or perhaps a baseball, targets that are struck with impact velocities of only a few kilometers per second.

Scientists have known for some time that the energy unleashed by an impacting meteor is far greater than that of an equivalent mass of even the most powerful chemical explosives. For this reason, they normally think of impacts between bodies moving at relative velocities of more than a few kilometers per second as explosions. An important conclusion from impact experiments is that disruption of a small body depends not only on the energy imparted by the impacting projectile, which is in turn a function of its mass and speed, but also on the strength and mass of the target. As the size of the target body increases, the importance of the inherent strength of its constituents diminishes.

The very existence of Stickney crater tells us that Phobos survived the impact (barely), but a slightly more energetic collision would certainly have blasted this tiny moon to bits. Proof that such disruptive collisions have actually occurred is provided by the existence of *asteroid families,* formed when asteroids, probably no larger than a few hundred kilometers in diameter, collided and broke apart into pieces still large enough to be observed with a telescope. Neither of the parents of such dysfunctional families apparently survived the mating to tell the tale. The siblings born of such violence are, naturally, all the same age, but they tend to come in assorted sizes and shapes. They can be recognized as members of the same family because they still occupy orbits that share certain common characteristics, such as their elliptical form, their inclination from the ecliptic plane, and the distance from the Sun at their farthest reaches. This is not at all the same thing as being bunched physically close together in space, and these asteroidal siblings may actually be far apart from each other at any particular time. Nevertheless, by analyzing the similarities in their orbits it is possible to recognize chunks of impact debris that may have once been parts of the same body.

Siblings in an asteroid family compositionally resemble one parent asteroid or the other, since all are actually pieces of either Mom or Dad rather than some genetic combination of the two. The composition (parentage) of each piece can be guessed by measuring its *spectra,* or in other words, by breaking the sunlight reflected off its surface into its component wavelengths. Some asteroid families appear to be pretty much spectrally uniform, implying either that Mom and Dad were alike and uniform in composition (that is, chondritic) or that chondritic Mom was much larger than Dad so that his progeny did not add much to the mass of the family. Other families are spectrally diverse, perhaps the result of a mixed marriage or the disruption of one or more parents that were already differentiated before impact.

Once formed, an asteroid family may persist for billions of years, but older families may have to contend with further complications. For example, some members of older families may be lost as they undergo subsequent collisions that grind them into particles too small to be observed, or the family may "adopt" unrelated interlopers that are knocked into similar orbits.

Another conclusion gained from disruptive impact experiments is that most of the smashed material occurs not as big fragments (that is, members of asteroid families that are observable with telescopes), but as chunks much too small to be seen from afar. The fact that so much material appears to be missing hampers our ability to reconstruct the parent asteroids for families. These tiny bits, however, might be flung far enough afield to cross the orbit of the Earth and perhaps eventually to fall as meteorites.

Of course, it is not necessary that a meteorite parent body be destroyed during a collision in order to remove pieces from its surface. Asteroids are really rather tiny and the energy required to accelerate fragments to their escape velocities is correspondingly small. Standing on Phobos, you could loft a rock the size of an orange into Mars orbit with a baseball toss. Nevertheless, at least in some cases meteorites have been torn from their parent asteroids as massive impacts obliterated them, rather than simply having been gradually chipped off asteroid surfaces by small impacts. As one example, let's consider an interesting class of meteorites made predominantly of iron metal.

Among meteoritic irons, the twisted and scorched masses of metal that museums are so fond of displaying, the IIIAB class is the most abundant type. (The byzantine system of classifying and naming iron meteorites is of little consequence to our present discussion, so I will not subject you to an explanation.) Metal meteorites such as the IIIAB irons

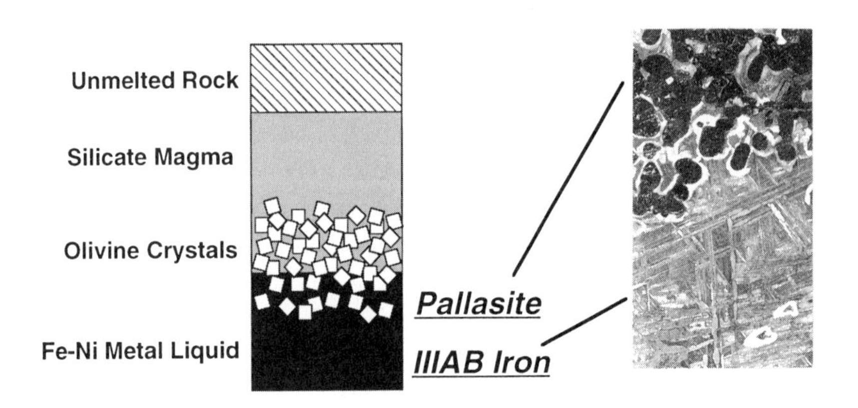

The sketch illustrates the formation of two types of meteorites in the interior of a strongly heated asteroid. The molten iron-nickel alloy segregated into a core, and olivine crystals in the overlying silicate magma sank and were pressed into the metallic liquid at the core-mantle boundary. Solidification of these materials yielded pallasites and IIIAB irons, represented in the photograph by olivine-rich and metal-rich areas of the Brenham, Kansas, pallasite. This sample has been etched with acid, so that the Widmanstatten pattern of intergrown metal plates is visible. (Smithsonian Institution photograph)

originally formed deep inside a strongly heated asteroid, as dense molten metal segregated into blobs that sank to the center. Based on their peculiar chemistry, we can recognize separate iron meteorites that fell to Earth at different times and places as having been derived from the same asteroid. Elements other than iron, such as nickel, iridium, and germanium, form metallic alloys with iron, and their concentrations in meteorites can serve as fingerprints for specific asteroids, including the one that once contained the IIIAB iron core.

Interestingly, another meteorite class, the pallasites, also contains metal with this same distinctive chemical fingerprint. The pallasites differ from the IIIAB irons in that they also contain another abundant mineral, olivine (as green crystals of gem quality, also known as peridot). The olivine crystals are thought to have been pressed into a mass of molten iron as they settled to the bottom of a chamber of silicate magma in the mantle overlying the IIIAB iron core, so we can consider the pallasites to be solidified samples from the core-mantle boundary of this asteroid. It appears that the world's meteorite collections contain a reasonably complete sampling of the heart of this particular asteroid.

To gauge just how difficult it might have been to extract samples of metal from this differentiated asteroid's interior, we must first determine how deeply they were buried and, since we know that they formed the core of this body, we can assume that depth of burial is roughly equivalent to the asteroid's radius. Its original size can be estimated from the rate at which pieces of the metallic core cooled. Once the molten metal and silicate magmas solidified, what remained was a metallic core encased in a rocky mantle that functioned something like a Thermos jug. The thicker this rocky blanket was (that is, the larger the asteroid was), the longer it could hold in its heat and the slower its core cooled. After the core was solid but before it was fully cooled, atoms of nickel in the solid metal did a curious thing. Initially distributed uniformly throughout the solid metal, the nickel atoms slowly rearranged themselves and segregated to form plates of a new, nickel-rich alloy. In meteorite samples, these plates now form a regular geometric array, called a *Widmanstatten pattern* (named for an Austrian count who first observed the structure in 1808). The pattern becomes visible when the metal is etched with acid, as shown in an accompanying figure. The speed at which the nickel atoms migrated through the solid metal was proportional to its temperature, and so indirectly to the rate at which it cooled. When the temperature finally reached a sufficiently low level, the nickel atoms stopped moving altogether. Measurement of the concentration of nickel at the edges and in the interiors of these plates reveals a frozen profile of diffusing nickel atoms, and from such analyses we can determine the cooling rate. The procedure for converting nickel profiles into cooling rates and, subsequently, into asteroid depths is rather complicated, but the bottom line is that IIIAB irons cooled at rates appropriate for the core region of a parent body at least several hundred kilometers in diameter, significantly larger than tiny Phobos but still of asteroidal dimensions.

But was this asteroid smashed into pieces all at once in a massive collision, or was its core slowly denuded and then sampled by many small impacts over some extended period of time? We can choose between these possibilities by making one further observation and one further measurement on these meteorites.

The observation is that the metal grains in most IIIAB specimens are very strongly deformed, actually riddled with microscopic dislocations that are characteristic of metal to which pressures three-quarters of a million times that of the Earth's atmosphere were applied instantaneously and uniformly. No other iron meteorite group shows this shock effect so pervasively. Small impacts are not likely to have produced such a high level of shock, nor would many separate impacts have affected such a high proportion of the samples to the same degree. The pervasiveness of this shock effect argues for a catastrophic impact.

The measurement we will make is a determination of the length of time the meteorites were exposed to *cosmic rays*. Meteorites orbiting in space are continuously bombarded by very tiny and very fast-moving particles emanating from the Sun and from elsewhere in the galaxy, and their effect is to produce a sunburn on whatever they touch. The sunburn is not visible; instead, it is a chemical change. This cosmic irradiation actually causes the transmutation of certain atoms near the skin of the orbiting meteoroid, a kind of alchemy that produces a variety of new radioactive isotopes. The cosmic sunburn halts when the meteorite falls to Earth, because the atmosphere shields out most cosmic rays. Some of the newly created isotopes are sufficiently different from the bulk of the meteorite that they can be recognized and analyzed. From measurement of the quantities of such isotopes in a meteorite, we can estimate how long it was exposed, or in other words, how long it existed as a small object in space. This length of time is called the *cosmic ray exposure age*. All the IIIAB irons have exposure ages that cluster around 700 million years. We would certainly expect to see a tight concentration of ages like this if these meteorites had been liberated from their parent asteroid all at once, but clustering of exposure ages would not occur if they had been gradually sampled during surface erosion by many small impacts over a long time. Taken together, the observation of strong and pervasive shock effects and the clustering of measured cosmic ray exposure ages seem to demand that a single catastrophic impact disrupted the IIIAB parent asteroid 700 million years ago.

FAMILY REUNION

My mother orchestrates family reunions, seemingly at every opportunity. Most of the relatives grumble but dutifully attend, and are amply rewarded with warm hugs, cold fried chicken, and rekindled memories. In some ways, asteroid families can be considered like people families. Like us, the members of asteroid families share a common inheritance, they sometimes break up, and they occasionally even have reunions.

I am leading up to the idea that sometimes fragments may actually reunite into a new object, but first let us examine some asteroid family members that, after disruption, merely remained in the same neighbor-

This photograph of asteroid Ida, taken fourteen minutes before the closest approach by the *Galileo* spacecraft, shows a tiny moonlet in orbit about the larger asteroid. (NASA)

hood. Double asteroids, called contact binaries, may be quite common, judging from a close look at asteroid Ida. In Greek mythology, Ida was a nymph who protected the infant Zeus from his violent father. It now appears that asteroid Ida, like its namesake, guards a baby. In 1994 the *Galileo* spacecraft photographed Ida and a tiny moonlet orbiting perhaps a hundred kilometers away from Ida's center. The moonlet is named Dactyl, after the Dactyli, supposedly Ida's children by Zeus. How did asteroid Ida come to be a baby-sitter? The capture of such a moonlet is a very improbable event; a close encounter would certainly deflect the smaller object, but it could not actually be captured unless a third force of some kind slowed it down. *Galileo* scientists have argued that these bodies are both siblings of an asteroid family, still together after the rest of the family dispersed. (Both, in fact, are members of the Koronis family that travels the main asteroid belt between Mars and Jupiter.) An impact that destroyed the ancestral Koronis asteroid may have shot Ida and its moonlet away in a single jet of debris, so that they had similar enough speeds and trajectories for the larger one to capture the baby.

The point is that fragments of bodies disrupted by impact do not always disassociate. In the case of Ida and Dactyl, several fragments remained in tight, intertwined orbits, but in other cases stronger gravitational attractions between siblings may actually cause them to reunite into a new object.

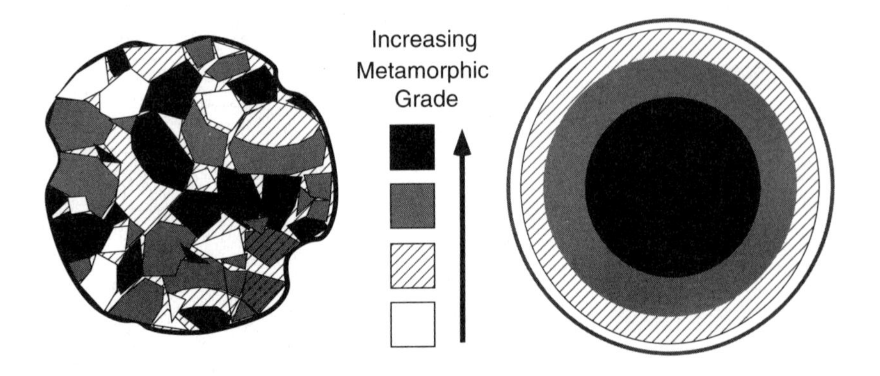

Rubble pile asteroids are objects reassembled from the broken debris of a body or bodies disrupted by impact. Chondrite asteroids were heated and metamorphosed, probably resulting in bodies with onion shell structures in which the most severely heated rocks (those at the highest metamorphic grade) were in the deep interior. Disruption and reassembly of an onion shell asteroid would produce a rubble pile, with rocks of different metamorphic grade distributed throughout the body.

We might describe an asteroid formed from many such reassembled fragments as a "rubble pile." Evidence for the actual existence of rubble pile asteroids comes from another class of meteorites, the *H* chondrites. (The *H* stands for "high iron," a chemical characteristic that distinguishes these meteorites from other chondrites.) These meteorites are mostly metamorphic rocks, stones that have been heated (but not to the point of melting, as they were in the IIIAB iron meteorite parent body) and recrystallized inside an asteroid. A body undergoing such metamorphism would presumably be hottest in its deep interior and progressively cooler near the surface, giving rise to a layered thermal structure that has been termed an "onion shell." Collisional destruction of an onion shell body, followed by reassembly of the pieces, would produce a three-dimensional jigsaw puzzle, a jumble of rocks that had been previously metamorphosed at different depths in an earlier asteroid.

In 1866 Cangas de Onís, a village in Spain, was peppered with meteoritic stones, a rainstorm of heavenly gravel from a large, *H*-chondrite meteor that broke apart during its transit through the atmosphere. Samples of this shower, collectively known as the Cangas de Onís meteorite, turned out to be rather peculiar. The meteorite specimens consist of broken, angular rock fragments mixed with fine particles of pulverized rock dust, now welded into a coherent rock. Geologists call such rocks *breccias*. Careful inspection shows that the angular rock fragments are actually pieces of chondrite that have been metamorphosed to different degrees. In effect, the Cangas de Onís meteorite breccia is a rubble pile in miniature.

How do you make a rubble pile that you could hold in your hand? To answer this question, we need to know how deeply buried were the various clasts of metamorphic rock before their incorporation into this breccia. By analyzing the nickel contents of metal alloys, which occur as tiny disseminated grains in the Cangas de Onís meteorite, it is possible to estimate their rates of cooling. (The procedure is another application of the metal cooling rate technique we earlier applied to the IIIAB iron meteorites.) If all of the metal grains in a meteorite cooled at the same rate, their compositions would plot along one of the cooling-rate contours shown in an accompanying figure. However, the metal grains in Cangas de Onís indicate wildly varying cooling rates, ranging from 1 degree to 1,000 degrees per 1 million years. Jeff Taylor and his colleagues at the University of Hawaii have estimated that the most slowly cooled metal grains in meteorites like Cangas de Onís had to have been buried at a depth of 100 kilometers in a parent asteroid that was perhaps 400 kilometers in diameter!

The Cangas de Onís breccia is actually a regolith breccia, meaning that the particles that comprise it once were soil covering the surface of the *H*-chondrite asteroid. The word *regolith* comes from Greek words meaning "stony blanket." So 100-kilometer-deep metal grains now reside in a compacted sample of soil taken from the uppermost surface of an asteroid. How did metamorphic rock fragments containing metal grains that once were deep inside this asteroid come to be on the surface? Impact, surely, but no impact is capable of excavating material from such a great depth in an asteroid without disrupting it. The existence of breccias like Cangas de Onís can only be explained if we accept that pieces of the shattered asteroid were reassembled into a new body, a rubble pile. In this way metamorphic rocks from throughout the body were embedded into the new surface layer.

Direct visual evidence for the existence of rubble piles may come from Miranda, a tiny inner moon circling the gas giant Uranus. The larger satellites of Uranus are all named for characters in Shakespearean plays. Following this tradition, discoverer Gerard Kuiper in 1948 named Miranda after the irascible, violent heroine of Shakespeare's *The Tempest*. Its name is fitting.

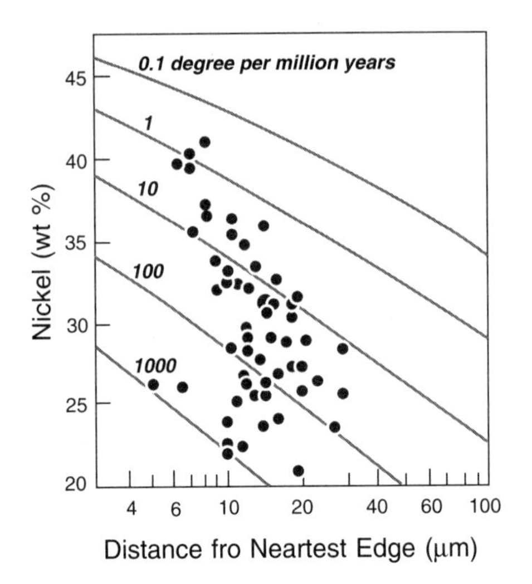

The Cangas de Onís, Spain, meteorite is a sample of compacted regolith from the surface of an asteroid. Metallographic cooling rates can be determined by measuring the migration of nickel. Metal grains in this meteorite cooled at rates varying from 1 to a 1,000 degrees per 1 million years (shown as contours). The slowest cooling rates indicate that the rock clasts that contain those metal grains were buried to a depth of perhaps 100 kilometers. Their incorporation into a surface soil sample indicates impact disruption followed by reassembly into a rubble pile asteroid. (Modified from C. V. Williams, A. E. Rubin, K. Keil, and A. San Miguel [1985], *Meteoritics,* vol. 20)

The *Voyager 2* spacecraft photographed the bizarre surface of Miranda in remarkable detail. Although *Voyager* scientists expected Miranda to be a bland ball of ice, it turned out to be rather exotic. This moon contains two strikingly different types of terrain, one heavily cratered and similar in appearance to the lunar surface, and the other decorated with alternating light and dark bands (actually ridges and troughs) sometimes folded into chevrons. Cutting through this jumbled montage of rock and ice are huge fault scarps and deep rifts that can be traced across the globe. And all this diversity occurs on a body that is only 560 kilometers in diameter, not much larger than the original *H*-chondrite asteroid and with barely enough gravity to pull itself into a sphere.

How did Miranda get this way? One explanation is that this moon consists of reassembled fragments from bodies that were shattered during a violent collision. The ancestral Miranda may have originally had a rocky core and an icy mantle, and interior pieces became intermixed

The bizarre appearance of Miranda, a moon of Uranus, suggests that it may be a rubble pile of reassembled pieces. (Jet Propulsion Laboratory)

with exterior ice on the surface of the reaccreted rubble pile. An alternate view is that the strange appearance of Miranda resulted from internal heating which caused incomplete separation of dense rock and ice; in this scenario, Miranda represents an arrested stage of planetary differentiation rather than a reassembled pile of planetary wreckage.

Other possible rubble piles include two of Saturn's moons, Janus and Epimetheus. These bodies skirt the outer edge of Saturn's rings, and their extremely low densities suggest that they may be loose aggregations of ring particles, which in turn may be the debris from an earlier satellite destroyed by impact. There is evidence that asteroid Ida, too, may be a rubble pile. Its density seems too low for compacted rock, implying that it may contain cavities, as expected for a body of jumbled rock.

SCARS, TILTS, AND SWAPPED ROCKS

Now let's turn to bodies significantly larger, more akin to planet Earth than to asteroids or miniature moons. Most people now accept the idea that the dimpled and acned faces of the Moon and of the other terrestrial planets have been ravaged by impacts. Among the cratered surfaces of worlds other than our own, one does not have to look very hard to find examples of massive collisions that, had the impactor been much larger or faster, might have torn both objects asunder. Such impacts left in their wake huge scars, called *multiring basins* because they are surrounded by several concentric sets of mountains.

The Imbrium basin is the most obtrusive scar on our closest neighbor, an impact structure that dominates much of the northern half of the lunar nearside. It is the large, dark splotch that figures prominently in the naked-eye visualization of the man in the Moon. G. K. Gilbert (the same geologist who conducted early impact experiments with clay) first described Imbrium as a collisional scar in 1893. Surrounding Imbrium Gilbert saw distinctive radial grooves, which he termed Imbrium sculpture, and which we now recognize as one of the hallmarks of impact craters. The basin has at least three rings of mountains, stretching the basin to a diameter of 1,140 kilometers. I say "at least three rings" because some of the Imbrium rings are exceedingly difficult to see. Many smaller craters are superimposed on the basin, and it was later flooded with basaltic lavas; both of these complications blurred the outlines of this structure. Material ejected from the impact was spread over much of the lunar surface. Because of its wide distribution this material, collectively called the Fra Mauro Formation, serves as an important geologic marker for understanding lunar geology.

At the antipode of the Imbrium basin lies a strangely furrowed terrain, thought to be an analog of the antipodal features of crater Stickney on Phobos, a region where seismic energy from the impact was focused after traveling around and through the body. Similar features occur at the antipodes of the Hellas basin on Mars and the Caloris basin on Mercury, suggesting that the impacts that produced the largest basins on these planets may have been nearly disruptive.

An even more awesome multiring basin is the Valhalla structure on Callisto, the second-largest moon of Jupiter. As imaged by the *Voyager* spacecraft, the bull's-eye Valhalla structure has twenty-five clearly visible concentric rings and perhaps hundreds of smaller rings. Its 4,000-kilometer diameter is comparable to the diameter of the satellite itself, which measures less than 5,000 kilometers. The entire continental United States could fit comfortably inside this ringed structure. Callisto is covered with a crust of ice, which clearly responded to impact much differently than did the rocky Moon. Ice flows more easily than stone, and Valhalla's rings look like monstrous frozen ripples on a pond.

All the terrestrial planets, except Earth, show huge multiring scars that testify to past cataclysmic impacts. But they also show another strange feature that may have resulted from mighty blows: most have

On the top is a painting of the Moon showing the Imbrium basin (upper left), as it appears today. On the bottom, it is compared with another rendition of the Moon as it must have appeared shortly after the Imbrium collision. The basin was later filled with dark lavas, which obscured the inner mountain rings. (Paintings by D. Davis, U.S. Geological Survey)

spin axes that are tilted relative to the ecliptic plane. Uranus, lying literally on its side, is the most dramatic example, but it is not alone. All of the planets except Mercury and Jupiter are inclined at angles of at least 24 degrees.

Venus is unusual in another way. Unlike every other planet in the solar system, Venus rotates slowly backward. A plausible explanation for this unique attribute is that its direction of spin was changed during an impact. If the impactor had been larger, had a different speed, or had approached at a different angle, perhaps Venus might have even been destroyed.

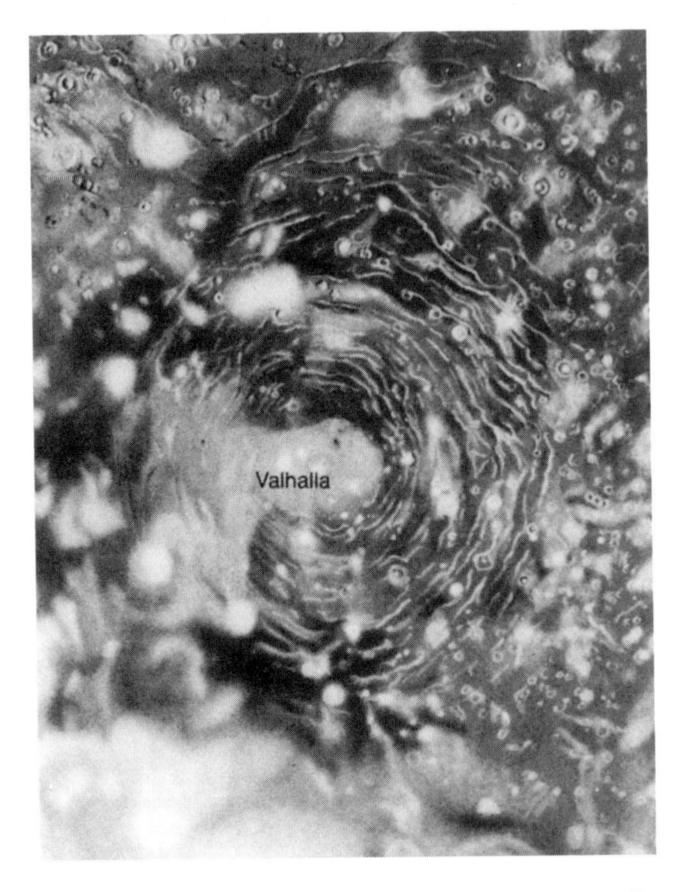

The gigantic Valhalla multiring basin forms a distinctive feature on Callisto, a moon of Jupiter. The numerous rings result from a massive impact into an icy crust. (Airbrush map, U.S. Geological Survey)

Large collisional events on moons and planets also appear to have had another, unexpected consequence. Perhaps it is no surprise that impacts between asteroids can dislodge rocks that eventually become meteorites. After all, the speed required for such rocks to escape a tiny asteroid's gravity field is correspondingly small. However, the escape velocities for planets would seem to be large enough to preclude planets from swapping rocks—or so we thought. It is now recognized that we have in our meteorite collections pieces of the Moon and probably Mars, rocks that must have been accelerated to speeds of several kilometers per second or greater during large impacts.

A NEW MOON

Mercury, Venus, Mars, the Moon, and even the outer planets and their moons all bear witness to collisions of massive proportion, pounding blows that scarred their faces or knocked them out of kilter. Now that we have explored the sometimes devastating effects of collisions on other worlds, it is time that we entertain the possibility of massive impacts on the early Earth. If the Earth's neighbors have been so catastrophically pummeled, how could our own world escape this fate? Obviously, it did not. In fact, Earth may have suffered the cruelest blow of all.

During planetary accretion, a hierarchy of planetesimals of various sizes were progressively assembled into ever-larger bodies. The distribution of object sizes is normally described by a power law, meaning that for every hundred Moon-sized bodies, there should have been about ten bodies the size of Mercury and perhaps one or two the size of Mars. Although there were many small bodies, most of the mass and energy resided in a few large ones. Therefore, a growing Earth embryo would have been battered by many small impacts (small only in a comparative sense), but most of the planet's actual growth would have occurred in a few catastrophic impact events.

The hypothetical collisional event we will now consider was gargantuan and had a truly remarkable consequence—the formation of the Moon. Competing suggestions about how the Moon formed have been around a long time. In the middle of the nineteenth century, scientists postulated that convulsive contractions or expansions of the Earth heaved off the mass of the Moon, leaving behind the Pacific Ocean basin. A few years later physicist George Darwin (second son of the biological evolutionist) postulated that a molten early Earth spun so rapidly that, when aided by tidal tugs of the Sun, it expelled the Moon. Darwin's idea eventually became widely accepted. The principal alternative to this idea was based on Laplace's nebular hypothesis. In this scheme the Moon condensed from a ring of matter spun off the Earth, just as the Earth itself condensed from a ring spun off the rotating solar nebula. Early in this century, scientists began to consider the possibility that the Moon formed elsewhere in the solar system and was captured by the Earth. An interesting variation on this theme was envisioned by physicist Hannes Alfven, who calculated that the Earth's gravity would have ripped away a portion of the Moon during an encounter close enough to effect lunar capture; he thus suggested that the crust of the

Earth was lunar material, a dramatic reversal of the century-old idea that the Moon was earthly stuff.

In 1975, planetary scientists William Hartmann and Donald Davis suggested that a massive impact might have thrown Earth material into orbit so that it later collected into the Moon. They did not specifically address a serious problem, however: fragments spalled off at less than the escape velocity would simply reaccrete to the Earth. The next year astrophysicists Alistair Cameron and William Ward offered a solution. They noted that an impact of the scale contemplated would vaporize much of the ejected matter, and expansion of this vapor would accelerate the gas and so allow it to escape the Earth's grasp. The cooling gas would condense to solid form and eventually be swept up into the Moon.

The evidence suggesting that the Moon formed during a massive collision is indirect, but such an origin seems to provide plausible explanations for some of the peculiarities of the Earth-Moon system when none of the other, historical mechanisms just considered can. The peculiarities to be explained are: No planet other than the Earth has a satellite so huge, relative to the size of the planet itself. The angular momentum of the Earth-Moon system, that is, the sum of the products of the masses of the two bodies times their angular velocities, is anomalously high compared with that of the other planets, and the orbit of the Moon is significantly tilted (a full 7 degrees) relative to that of the Earth. The bulk density of the Moon is much lower than that of the Earth or the other terrestrial planets, implying that the Moon contains much less metallic core material. Finally, the Moon is bone dry, with nary a drop of water, and other volatile elements are strongly depleted as well.

Before we see how a gigantic impact might have resulted in an Earth and Moon with these unusual properties, let's review the failings of other explanations. Darwin's hypothesis that the Moon was flung out of the mantle of a rapidly spinning Earth might explain the Moon's lack of core material, but the spin rate to achieve fission is now thought to have been implausibly high. Laplace's proposal that the Earth and Moon accreted in place as a double planet necessitates that both bodies were made from basically the same accreting materials; thus, there is no reason to expect these siblings to have different compositions and bulk densities, nor for them to have different orbital tilts. Capture of the Moon by the Earth is exceedingly improbable on dynamical grounds. All of these ideas stumble on their failure to account for the high angular momentum of the Earth-Moon system and, because they entail processes that would presumably have been common in the early solar system had they occurred at all, on the absence of similar satellites around the other terrestrial planets.

Many of these problems appear to be resolved if the Moon was produced during a massive collision. An impact of this magnitude might have been an unusual event, so the other planets would not be expected to have similarly large satellites. The collision itself would have imparted enough energy to account for the high angular momentum, and there is little problem in forming a moon with an orbit tilted from the ecliptic in such a scenario. The vaporized and recondensed rock that formed the Moon was devoid of water and depleted in other volatile elements, because the more volatile materials never fully condensed.

The compositional differences between Earth and Moon, reflected in their different bulk densities, can be understood from models of giant impacts. Models describing the outcome of a gigantic, Moon-forming impact require powerful supercomputers. One such study assumed that both colliding bodies were already differentiated with metal cores and rocky mantles at the time of impact. In this simulation, the masses of the Earth and the impactor were divided into about 3,000 particles, which were free to move about independently in reponse to gravity and pressure forces exerted on them. The resulting trajectories of the particles, as well as their temperatures and initial compositions, were tracked during the ensuing twenty-four hours. In repeated simulations, the investigators varied the size of the impactor, its speed, and other factors. The optimum model involved an impactor with a mass about one-seventh that of the Earth (which is larger than the planet Mars) with an impact velocity of 5 kilometers per second. There is obviously no unique solution to this kind of exercise, but some simulations resulted in Earth-Moon systems broadly similar to the real thing. For example, in one run the impactor was disrupted within minutes of impact and most of it was lofted into Earth orbit. Within four hours, the metal core material from the impactor had plummeted back to Earth. The rocky material remained in orbit and accreted to form a Moon-like lump within about a day.

The giant impact origin for the Moon is really another example of planetary disruption followed by reassembly, though in this case the result was two bodies rather than one. Whether the Earth and Moon can be termed rubble piles is a semantic argument, but the principle is

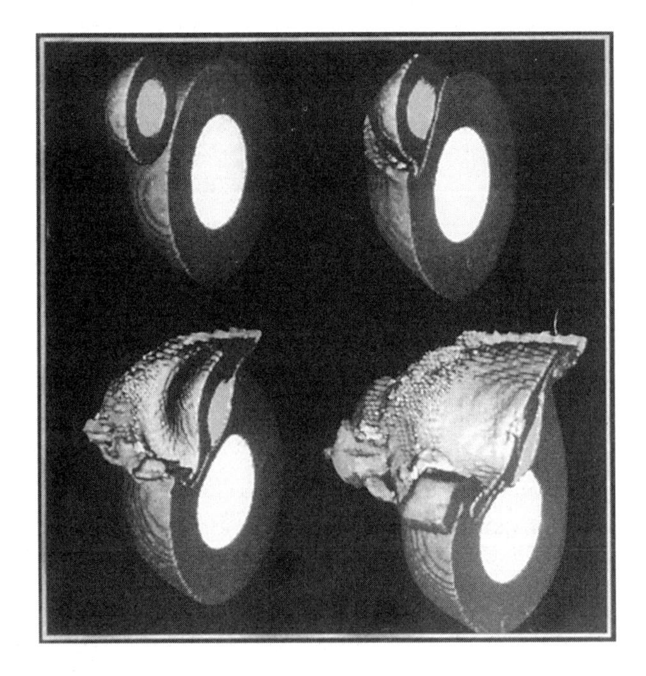

These images, from supercomputer models, illustrate the earliest stages of a gargantuan collision in which a Mars-sized impactor collides with the Earth. Both bodies are differentiated with cores and mantles at the time of impact. In succeeding frames, the core of the impactor merges with the core of the Earth, and much of the impactor's mantle is ejected into Earth orbit, where it reaccretes to form the Moon. (Computer simulations by H. J. Melosh, University of Arizona)

the same as we saw in asteroids: planets can suffer catastrophic impacts that at least partly disrupt them, and the dislodged materials can be reassembled into new objects.

CONSEQUENCES FOR THE PUMMELED EARTH

The collision of a Mars-sized impactor with the Earth must have transferred a prodigious amount of energy to our planet over a time interval measured in minutes. It seems an inescapable conclusion that a large portion of the battered Earth must have melted, forming a global magma ocean. The heat for such a global meltdown came from the kinetic energy of the impactor. The total energy available from the collision of a Mars-sized impactor was enough to raise the temperature of the Earth by about 10,000 degrees. The temperature of the Earth did not reach this extreme value, however; the efficiency with which energy of motion of an impactor is transformed to heat is poorly known, but estimates are no higher than 50 percent. Reasonable calculations suggest that the Earth's mantle may have been heated to temperatures of at least 3,000 degrees, though, well above the temperature at which it would melt.

Although the total *amount* of energy in such a collision is staggering, the *distribution* of energy within the Earth is equally important. Infalling material that consists of relatively small particles generates heat only in a surface layer of the planet, so that the body may become hotter on the outside than the inside. In contrast, huge projectiles release their energy initially as shock waves that propagate through the target planet, and as the shock waves decay they deposit their energy as heat in the deep interior. Deep melting should have been widespread within the Earth, at least in the hemisphere in which the impact occurred.

If the impactor had an iron core, it would likely have sunk through the molten mantle and merged with the Earth's core, thereby releasing even more gravitational potential energy. If the core of the impactor was, say, 30 percent of its mass, the energy released by sinking through the mantle is about a tenth of that for the impact itself, but still a considerable amount. When the cores merged, this energy was also converted to heat, which was applied to the bottom of the mantle, causing further melting.

The staggering temperature achieved just a few minutes into the impact event vaporized a considerable amount of rock. This vapor was expelled backward from the impact site, and most escaped and went into Earth orbit. However, some was probably retained and formed a transient atmosphere of searing silicate gas that covered and eventually rained down on the hapless planet's surface.

Another result of this massive impact is that the bulk composition of the Earth was actually changed. Some mantle material was lost during impact, and virtually the entire core of the impactor was added to the Earth. Our planet was probably not the only one to suffer such a consequence. Mercury is a peculiar planet, because it has such a very high bulk density, compared to the Earth and other terrestrial planets. Its metallic core must be huge, relative to its rocky mantle and crust. For many years scientists wrestled to explain this compositional difference from its neighbors, but the recognition that disruptive impacts may have played a role in the evolution of the Earth leads to another possibility. Perhaps in the distant past, Mercury was struck by a massive projectile, a glancing yet devastating blow that stripped off a significant fraction of its rocky exterior. What was left behind, again pulled into spherical shape by gravity, was a planet with too much core, another world whose very composition was altered by impact.

THE IMPORTANCE OF A CHANCE EVENT

In July of 1994, an astounding series of events took place. The world anxiously watched as, every few hours, a hurtling chunk of comet plunged into the atmosphere of Jupiter. All of the twenty-odd fragments, collectively called Comet Shoemaker-Levy 9 after its discoverers, were once part of the same object, now dismembered and strung out along the same orbit. This cometary train, glistening like a string of pearls, had been first glimpsed only a few months before its fateful showdown at Jupiter, and rather quickly scientists had predicted that the fragments were on a collision course with the giant planet. The prediction was right on target, as was the comet. The arrival of each piece was punctuated with an explosion clearly visible from Earth, a searing conflagration that mushroomed as each icy mass incinerated itself. When each fragment slammed at 60 kilometers per second into the dense atmosphere, its immense kinetic energy was transformed into heat, producing a superheated fireball that was ejected back through the tunnel the fragment had made a few seconds earlier. The residues from these explosions are now huge black eyes on the face of Jupiter, some of which have stretched out to form dark ribbons.

Although this impact event was of considerable scientific import, it especially piqued public curiosity and interest. Photographs of each collision made the evening television newscasts and were posted on the Internet. This was possibly the most open scientific endeavor in history. The face of the largest planet in the solar system was changed before our very eyes. And for the very first time, most of humanity came to fully appreciate the fact that we ourselves live on the flanks of a target, a world subject to catastrophe by random celestial marksmanship.

That realization was a surprise to many, but it should not have been. One of the great truths revealed by the last few decades of planetary exploration is that collisions between bodies of all sizes are relatively commonplace, at least in geologic terms, and were even more frequent in the early solar system. The effects of these crashes have been profound. In many cases, the targets have been scarred and their orientations or spins have been changed. In extreme cases, the targets and impactors were actually disrupted, and sometimes the debris was reassembled into new objects.

Had the Earth and an impactor the size of Mars not happened so long ago to occupy the same place at the same time, we could not now gaze upon the shining Moon—our planet would be alone in this corner of the solar system. And without a nearby Moon, this would be a very different place. The restless ocean tides would hardly slosh back and forth, driven only by tugs from the faraway Sun. Our nights would be terrifyingly black, and hundreds of love songs would have to be rewritten. The course of human history would certainly have changed drastically without a Moon to worship or wonder about; even the monthly human menstrual cycle might have evolved to some different rhythm.

Perhaps it is even possible that, without a Moon, the Earth would have been an inhospitable platform for life at all. French physicists modeling the stability of the tilt of the Earth's spin axis (the technical term for this is *obliquity*) have shown that our planet should have experienced dramatic and chaotic changes in obliquity over its lifetime, as have the other terrestrial planets. However, the orientation of the Earth's tilt is apparently locked into its present value (varying a paltry 1.3 degrees from the mean value of 23.3 degrees) by the gravity of its large satellite. Variations in obliquity are important because they are thought to cause planetary ice ages and warm spells. The Moon may be, in this sense, a climate regulator. If there were no Moon, or if it were much smaller, the Earth's obliquity would reach values as high as 85 degrees, with probably disastrous consequences for any life forms trying to weather the resulting climate oscillations. Life on Earth may owe its existence, in part, to a catastrophic planetary impact eons ago.

Some Suggested Readings

Chapman, C. R., and Morrison, David (1989). *Cosmic Catastrophes.* Plenum, New York. *An excellent review of the emerging scientific debate on catastrophisms; this popular book includes a chapter on the giant impact origin of the Moon.*

Hamblin, W. K., and Christiansen, E. H. (1990). *Exploring the Planets.* Macmillan, New York. *Many of the chapters of this excellent elementary textbook show crisp photographs of huge impact scars on the surfaces of planets and satellites.*

McSween, H. Y., Jr. (1993). *Stardust to Planets: A Geological Tour of the Solar System.* St. Martin's, New York. *Chapter 6 of this book describes the historical recognition of multiring basins as impact structures. If you are not enjoying the present book, you won't like this one either.*

Melosh, H. J. (1989). *Impact Cratering: A Geologic Process.* Oxford University Press, New York. *For those who can handle a more rigorous, physics-based treatment of impact phenomena; Chapter 12 discusses the role of massive collisions in planetary evolution.*

Steel, Duncan (1995). *Rogue Asteroids and Doomsday Comets.* Wiley, New York. *A popularized treatment of the potential hazards posed by impacts for the Earth's biosphere; the book also unfortunately includes a number of controversial or unsupportable ideas about the origin of these impactors.*

Taylor, S. R. (1987). The origin of the Moon. *American Scientist,* vol. 75, pp. 468–77. *An article that nicely summarizes current thinking about the Moon's origin by collision of the Earth with a Mars-sized impactor.*

Taylor, S. R. (1992). *Solar System Exploration: A New Perspective.* Cambridge University Press, New York. *This wonderfully crafted book literally celebrates catastrophic collisions as events that shaped the solar system.*

Rock of Ages

A Chronology of Creation

MULE RIDE OF EONS

The scenic vista afforded by the Grand Canyon is actually carved out of time as much as hard rock. In the space of little more than a vertical kilometer, the Colorado River has dissected one and a half billion years of Earth's history, a full third of the planet's total existence. In the summer of 1994, I persuaded my family to take a close look at this incredible temporal record via a mule ride from the canyon rim to the bottom. My wife was not particularly enthusiastic about this adventure, as she has this thing about heights. The mules that troop the Grand Canyon are trained to walk along the outside edge of the winding trail, so as to give hikers the inside passage. The assurance of the National Park Service that her mule was surefooted offered little consolation as she lurched, white-knuckled, along the precipice over the yawning abyss. Before starting out, our guide had suggested that anyone uncomfortable with heights should watch the adjacent canyon wall, rather than peer over the ledge, so I think she examined more geology during the trip than I did. When we finally reemerged that afternoon at the canyon lip, dusty, parched, sunburned, and relieved, our leg muscles were so cramped from six hours on muleback that we could barely hobble back to the lodge. But time travelers must expect some hardships.

The photogenic purple, red, and tan layers that comprise the bulk of the Grand Canyon walls—the Coconino and Tapeats sandstones; the Kaibab, Toroweap, Redwall, and Muav limestones; the Hermit, Supai, and Bright Angel shales—are all sedimentary rocks deposited in a succession of long-gone seas. Some of these formations contain fossilized remnants of unfamiliar creatures that inhabited the waters in those days. These rock layers, or *strata,* are piled up, one atop another like the pages of a book, in the same sequence in which they were deposited. Near the bottom, the canyon walls exhibit a ragged line, the trace of an ancient erosional surface. It is impossible to know what thickness of rock must have been eroded away when this surface was once dry ground. This *unconformity,* as geologists call it, represents missing time, even more time than it took to deposit all the rock layers above it. (This is not the only unconformity in the Grand Canyon sequence, but it is the most noticeable and encompasses the greatest amount of time.) The partially eroded rock unit situated just below this unconformity, the very base of the Grand Canyon itself, is the Vishnu schist. At one time the Vishnu basement, too, consisted of sediments, but about 1.4 billion years ago they were metamorphosed by heat and pressure and intruded by hot granitic magmas, so that their original form is now unrecognizable. More than 800 million years elapsed between the time that the Vishnu sediments were transformed into schist and the time that the sands of the immediately overlying Tapeats formation were deposited, a hiatus of almost unfathomable duration squashed into a thin, wavy line.

As incredible as this marvelous window on the Earth's history is, the Grand Canyon tells us nothing about the creation of our planet or of the events that preceded or shortly followed it. No rocks, none anywhere on the globe, are preserved from that opening epoch. We can only travel part of the way back in time to the planet's birth. It is possible, nonetheless, to fix the time of our beginning, and that we will do in this chapter.

But this chapter is about more than merely determining the chronology of our planet's creation. Geologist Brent Dalrymple, in the preface to *The Age of the Earth,* noted that a colleague who learned that he was writing a book by that title expressed the opinion that it would be a short book, because the answer could be expressed as a single number. The colleague missed the point. The tortuous path by which the number was arrived at is more interesting than the number itself, and the story of this quest reveals much about the philosophical underpinnings and strategies of science.

The Vishnu schist, seen in the foreground of this photograph near the bottom of the Grand Canyon, is an ancient metamorphic rock intruded by granitic magma. A former erosional surface, separating the Vishnu schist from the overlying sedimentary layers, is seen at the upper right; this curved unconformity represents missing time in this stratigraphic sequence.

SEDIMENTAL JOURNEY

Our concept of an Earth of finite age is a modern idea. The ancient Greeks rejected the notion of finite time altogether, instead favoring the idea of a cyclic universe that was regenerated again and again. For them, as well as for the ancient Hindu and Babylonian cultures, time was closed, like a circle. There was no beginning and no end, only endless cycles.

Hebrew and Christian traditions viewed time differently, although in medieval times they mostly ignored the chronology of creation. With the Reformation, however, the church's sole authority to interpret

Scripture was challenged and, rather than face the specter of nonconformist thought, the church began insisting that the Bible be accepted word for word. In 1654, James Ussher, Archbishop of Armagh in Ireland, literally summed the biblical lineages of the Old and New Testaments and concluded that the world must have been created in 4004 B.C. Armed with more vivid imaginations, other biblical authorities soon refined this estimate to a more precise time: October 26, at 9:00 in the morning; apparently the Creator was not an early riser. Ussher was not the first to use the Bible as an hourglass, only the most remembered, and other scholars before him had also arrived at similar ages for the Earth. The prominence of Ussher's date is probably due to its inclusion as a marginal note in the English Bible for over two centuries (it was finally removed in 1900). This biblical chronology went virtually unchallenged until the French naturalist Comte de Buffon, in 1749, attempted to determine the age of the Earth by experiment and scientific reasoning. His preferred time of creation, some 75,000 years ago, was based on his understanding of the Earth's internal heat and rate of cooling, and it broke sharply with established doctrine.

The primary challenge to the biblical idea that the world was young, however, came from a revolutionary interpretation of the geologic record. In 1785 James Hutton, a Scottish gentleman-farmer and physician, inaugurated this new scientific perspective. By this time scientists already recognized that geologic strata, like those so beautifully exposed in the Grand Canyon, were sediments that had been laid down in water, and it took only a small leap of faith to make a connection with the deluge described in Genesis. Hutton had spent years examining geologic formations in his native land, and he became convinced that there was no need to invoke the Noachian flood or, for that matter, any such catastrophe, to produce stratigraphic layers. Instead, he argued that the normal agents of geologic change, the same slow processes of erosion and deposition that occurred in his day, if given sufficient time, could account for all he saw. Of course, the key here was sufficient time, by which Hutton meant a lot of time. In his own words, "We find no vestige of a beginning, no prospect of an end."

A more pressing argument for lots of time, however, came from the fledgling science of paleontology. Fossils provide a means of correlating the strata that contain them from one place to another, in effect demonstrating that a particular limestone in France and perhaps a shale in America were deposited approximately at the same time. In the eighteenth century the study of fossils began in earnest. Correlation of strata using this new key allowed some of the time gaps represented by unconformities to be filled, further substantiating the idea that the worldwide succession of layered sedimentary rocks could not have been deposited at a hurried pace. In 1830 British geologist Charles Lyell published his *Principles of Geology,* in which he marshaled stratigraphic and paleontologic arguments to demonstrate, far more convincingly than had Hutton, that the forces at work on and within the Earth had remained constant in kind and degree throughout the ages. If the vast forces and dramatic events required by the catastrophists were to be rejected, something had to replace them. For Lyell, like Hutton, this something was time. It was simply not possible to compress all of geologic time into a few thousand years.

Although Lyell did not attempt to estimate the exact age of the Earth, others certainly did. If the geologists' focus at that time was stratigraphy, the most logical way for them to measure time was to quantify the rate at which sediments accumulated. In 1893 Charles Walcott (later to become secretary of the Smithsonian Institution) published the most thorough study of this kind; by summing the thicknesses of strata of various ages, he guessed that the whole of geologic time encompassed some 55 million years. In 1899 John Joly, an Irish mineralogist, reasoned that rivers, in addition to carrying sediments, must deliver salt from the continents to the sea. Assuming that no salt was ever lost from the oceans, he calculated that 90 million years were required to account for their present salinity. The ages derived by both Walcott and Joly were strict applications of Lyell's uniformitarian concept, although we now know that the methods on which both ages were based were flawed (Walcott's assumption that deposition occurred at a uniform rate and Joly's premise that nothing ever left the ocean were wrong). But even the great expanses of time implied by these estimates were not enough for naturalist Charles Darwin. In a way, Darwin was also Lyell's disciple in accepting the immensity of geologic time, and *Principles of Geology* had profoundly influenced his views. But Darwin raced past Lyell's idea that the world was in a state of dynamic balance, asserting that life evolved in a biologic progression that must have required hundreds of millions of years to accomplish. Many other students of the Earth soon followed his lead in advocating an old planet.

THE PHYSICS OF TIME

In the late nineteenth century, British physicist William Thomson (later Lord Kelvin) took on geologists by arguing that the Earth was young, and for a time he won. This was a highly visible battle, followed by scientist and layman alike:

> It takes a long time to prepare a world for man, such a thing is not done in a day. Some of the great scientists, carefully ciphering the evidences furnished by geology, have arrived at the conviction that our world is prodigiously old, and they may be right, but Lord Kelvin is not of their opinion. He takes the cautious, conservative view, in order to be on the safe side, and feels sure it is not so old as they think. As Lord Kelvin is the highest authority in science now living, I think we must yield to him and accept his view. He does not concede that the world is more than a hundred million years old.
>
> —Mark Twain, *Letters from the Earth*

Kelvin was provoked by Lyell's proposal of a steady-state Earth, because to him it smacked of a perpetual motion machine, a clear violation of the basic laws of physics. Likewise he rejected Darwin's hypothesis of an evolving Earth, as it did not accord with his preference for design in nature. Kelvin envisioned the Earth as an engine driven by heat, and heat was something he knew a lot about. He had already devised the absolute temperature scale (its basic unit, the kelvin, still bears his name). More important, he had established a fundamental relationship between heat and energy, now known as the second law of thermodynamics. Simply stated, whenever energy is converted from one form to another, some fraction of that energy is irretrievably transformed to heat. As a result, the total amount of energy in the universe remains constant, but every energy transformation (that is, every geologic process) must finally result in the net loss of energy available for useful work. So as a consequence, the universe must be running down.

This continuous dissipation of energy clearly ran counter to the geologists' notion of uniformity, but Kelvin was convinced that nature could not operate in apparent violation of thermodynamics. He could think of but one ultimate and universal energy source—gravity. For the Earth, he envisioned that its original energy, in fact its only energy, was derived from the gravitational accretion that had brought it into being.

William Thomson, Lord Kelvin, the most influential scientist of his day and a formidable foe of the idea of an ancient Earth.

Thus he saw the planet as initially very hot but in a constant state of cooling.

To fix the age of the Earth, Kelvin had merely to calculate the time necessary for the Earth to cool from its assumed primitive state to its present condition. As independent checks, he also calculated the age of the Sun, assuming that its heat was due to the continued contraction of the matter that formed it, and another age of the Earth based on the frictional braking of its rate of rotation over time. Both of these calculations, too, assumed that gravitation was the only possible source of energy. Kelvin initially settled on an age of 100 million years for creation, although over the next three decades he systematically revised the estimate downward to 24 million years.

Many Victorian geologists and biologists were uncomfortable with Kelvin's estimate but did not wish to violate the laws of physics, especially as championed by such a prominent and influential scientist. Backbone came from an unexpected quarter. John Perry, one of Kelvin's former assistants, had the effrontery to question the façade of mathematical certainty in Kelvin's arguments by pointing out the unfounded preconditions and assumptions undergirding his mentor's calculations. Perry demonstrated that the cooling history of the Earth would have been dramatically longer if some of Kelvin's simplifying assumptions were omitted. The accumulating evidence of geology and paleontology had already convinced many scientists that something was amiss, and Perry's work showed them what specifically might be questioned in the physics that so haunted the notion of an ancient Earth. It also fostered a growing confidence among the dissenters. Thomas Chamberlain, a geology professor at the University of Chicago, soon pronounced that if so brief a history was prescribed by physics, then physics must be wrong. (Of course it wasn't physics that was wrong, merely its leading spokesperson.) Chamberlain thought that some other unknown source of heat must power the Earth's engine, some almost inexhaustible fuel to match the planet's great age: "What the internal constitution of the atoms may be is yet an open question. It is not improbable that they are complex organizations and the seats of enormous energies." Chamberlain's prophetic statement was shortly to be proved correct. Ironically, Lord Kelvin, too, was shown to have been right when he derided the age estimates of the geologists, but wrong when he thought them to be too long.

RADIOACTIVITY TO THE RESCUE

In 1903 Ernest Rutherford and Frederick Soddy, then both faculty members at McGill University in Canada, demonstrated that radioactive atoms disintegrate spontaneously, regardless of whether the atoms are isolated or chemically combined with other elements. Even more suprising, they discovered that radioactive substances, as they decay, release heat. The implications that this finding held for understanding the age of the Earth were not lost on the discoverers, nor for that matter on anyone else. If the Earth possesses its own internal heat source, in the form of buried radioactive elements, then the basic premise of Kelvin's chronology—that an initially hot planet was inexorably cooling—must be faulty, as must be his 24-million-year age.

The next year Rutherford was invited to England to lecture on his discoveries before the Royal Society, Kelvin's very backyard. Rutherford later described the experience this way:

> I came into the room, which was half dark, and presently spotted Lord Kelvin in the audience and realized that I was in for trouble at the last part of the speech dealing with the age of the Earth, where my views conflicted with his. To my relief, Kelvin fell fast asleep, but as I came to the important point, I saw the old bird sit up, open an eye and cock a baleful glance at me! Then a sudden inspiration came, and I said Lord Kelvin had limited the age of the Earth, *provided no new source of heat was discovered.* That prophetic utterance refers to what we are now considering tonight, radium. Behold! The old boy beamed upon me.
>
> —A. S. Eve, *Rutherford* (1939)

Despite Kelvin's beatific smile on that evening, he never retracted his published age for the Earth, nor did he cease debating those who advocated radioactivity as an important heat source. He died an unregenerate foe of this idea, but his cause was futile.

At first, the discovery of radioactivity seemed to lay waste to the most quantitative method available for calculating time, but shortly it was to give rise to an even more rigorous clock. In studying radioactive thorium, Rutherford and Soddy determined that the proportion of radioactive atoms that disintegrate in a given time interval is an unvarying constant. Leapfrogging from this insight Rutherford, in a 1905 lecture at Yale University, then made the bold suggestion that radioactive decay was a potential chronometer. Because the quantity of a radioactive element decreases in a systematic manner with time, one could measure the passage of time by analyzing this change in abundance.

The specific timepiece that Rutherford favored was a radioactive decay process that produced helium, since he had observed its formation from thorium and uranium in his laboratory. A serious problem stood in the way of his applying this tool, although he found an ingenious solution. To understand this problem, let us take as an analogy an hourglass. At my house we sometimes play games that require timed responses, so the games come equipped with small hourglasses (actually, "minuteglasses" would be more accurate). Elapsed time is assessed by noting when the sand grains filter from the upper compartment of the hourglass to the lower, assuming of course that all the sand started out in the upper compartment as the hourglass was turned over. However, in the excitement of the game sometimes a player will hurriedly upend the glass before all the sand has had a chance to collect at one end. In principle, we could actually determine the elapsed time anyway, provided that we knew exactly how much sand was in each compartment at the start and the finish (but of course we don't). This is also the situation for radioactive atoms; the transformation from one element to another is incomplete at the time the measurement is made. Rutherford recognized that each uranium atom decay (think of uranium atoms as sand grains in the upper compartment) produces eight helium atoms (sand in the lower compartment), and so by measuring the helium he could calculate how many uranium atoms had decayed. Since he could measure directly the amount of uranium still present, he could then determine the proportion by which the uranium had decreased, and thus the age of the material that contained these atoms. Applying this radioactive hourglass to several minerals rich in radioactive elements, Rutherford obtained ages of about 500 million years. He recognized a potential problem with this technique, however. The decay product, helium, is a gas that can leak out of crystals, and so the age determined in this way may be merely a minimum age. It turns out that this problem is severe enough that this particular hourglass eventually had to be abandoned.

A short time before Rutherford made his startling proposal that the helium produced by radioactive decay of uranium could serve as a clock, a Yale University chemist by the name of Bertram Boltwood began analyzing uranium minerals. Boltwood found that such minerals invariably contained lead, and so he surmised that lead, in addition to helium, must be a product of radioactive decay. (This was before isotopes had been discovered. We now understand that uranium and thorium isotopes decay in a succession of steps, with helium produced all along the path, which ultimately ends at stable isotopes of lead.) Aware of Boltwood's finding, Rutherford suggested that age calculations made on the basis of the amount of lead in radioactive minerals might be more accurate than those based on helium, because the lead created in a compact mineral would have less possibility of escape.

Boltwood soon put this idea to the test. By 1907, he had analyzed lead in forty-three uranium-bearing minerals from ten localities. Eliciting the aid of geologists, he also compiled a list of the *relative* ages of these minerals, based on the order of appearance of the strata from which they were collected. By comparing these data, Boltwood found that different types of uranium minerals of the same relative age had a constant ratio

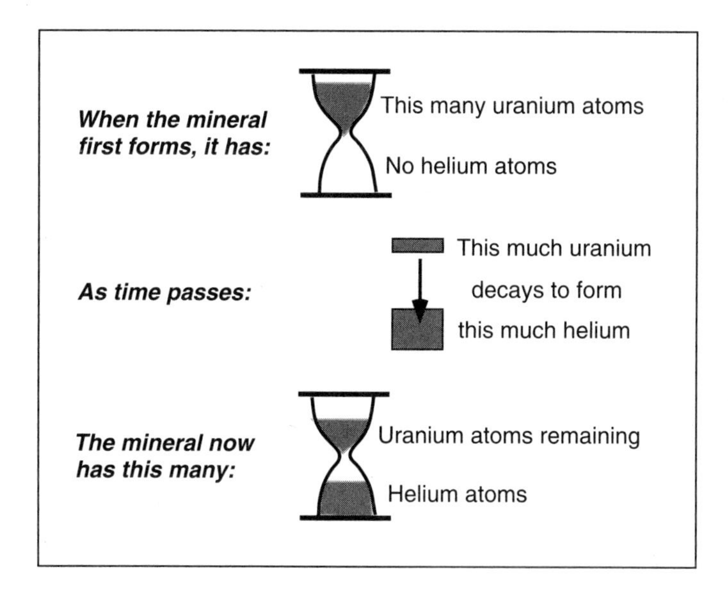

Hourglasses illustrate the workings of the uranium-helium chronometer. Helium is not initially present when the mineral forms, but it is produced by radioactive decay of uranium atoms over time. By knowing the number of atoms of helium that result from each uranium decay, it is possible to calculate the proportion of uranium that has disintegrated. Then, knowing the rate at which the decay occurs, we can determine the age of the mineral.

of uranium to lead, and that stratigraphically older minerals contained higher proportions of lead. The logic that lead was the product of uranium decay seemed justified.

Boltwood was now in a position to calculate the minerals' *absolute* ages, although his real interest lay in identifying the elements that formed during radioactive disintegration. Prodded by Rutherford, with whom he corresponded extensively, Boltwood calculated ages for his minerals that ranged from 400 million to 2.2 billion years. Part of Boltwood's reticence to take this step may have been that he recognized the ages were very uncertain, because he had to guess at the rate at which uranium decays. We now know, by taking into account the precisely determined rate of uranium decay and other factors, that his ages were in fact too long, sometimes by hundreds of millions of years. Imperfect as they were, though, these highly experimental ages represented a great leap in the understanding of geologic time. They rendered obsolete all earlier methods for estimating the age of the Earth, methods that were really based more on scientific intuition than physical principles. At the same time, they hatched the radical concept that the Earth might be billions, rather than millions, of years old.

LEAD AND THE AGE OF THE EARTH

As elements go, lead doesn't get much respect. It is mostly notable for its great mass, prompting its common use in phrases like "heavy as lead" and more pejorative terms like "leaden prose" (which I hope will not be applied to this book). It also has a reputation as a substance of little value, as in "converting lead into gold." However, lead has another, salutary aspect, often overlooked: it has the best memory in the Periodic Table, an unusual trait that makes it the most precise timepiece on Earth.

Arthur Holmes was intrigued by the idea that geologic time could be quantified by measuring lead. Initially skeptical that Boltwood's dating method would work, this British physicist soon convinced himself that radioactivity could provide geology with the precise clock that it required. Thereupon, he abandoned physics for geology and began a career devoted to quantifying geologic time. Holmes, more than anyone else, was responsible for the mature development of this subject and its acceptance by the broader scientific community.

The ages determined by Boltwood and others gave only an indirect constraint on the age of the Earth: it must be at least as old as the oldest dated mineral. Holmes wanted to do better. He believed that all lead was produced by radioactive decay, so that when the Earth originally formed it must have contained little or no lead. By estimating the amount of lead and uranium in the Earth's crust, he then arrived at an age somewhere between 1.6 and 3.0 billion years. Holmes actually favored an age closer to 1.6 billion, noting that some radioactive elements may have existed in the Sun and there they generated some lead before the Earth was born. His estimated age, published in 1927, remained the accepted value until the Second World War.

Further progress only came on the heels of a more fundamental understanding of how lead's memory works. What made this new understanding possible was a revolutionary instrument, the mass spectrometer, which could actually weigh atoms and separate them from one another by mass. Invented by English physicist F. W. Aston in 1919, this instrument was redesigned in the 1930s into a tool of high precision by Alfred Nier of the University of Minnesota. During subsequent years, the discoveries that uranium consisted of two isotopes (with mass numbers of

235 and 238) and lead had four (with mass numbers of 204, 206, 207, and 208) revealed that the uranium decay scheme must be much more complex than Boltwood had envisioned. Holmes, along with several other scientists working independently, finally decoded this scheme. Naturally occurring uranium is a mixture of both radioactive isotopes, each of which disintegrates at a different rate. Uranium-235 decays via a complicated chain of transformations into other radioactive isotopes, until finally reaching stable lead-207. Uranium-238, which decays twenty times slower than uranium-235, descends a similar staircase of unstable isotopes, ultimately arriving at lead-206. Lead-208 forms from the decay of a different element, thorium, and lead-204 is not radiogenic, being the result of no known decay process.

Natural lead is a mixture of all four isotopes, one eternal and three created over time by radioactive decay. This was a major source of error in previous age determinations, which had assumed that all lead had but one radioactive source. Three uranium decay schemes mean that there are actually three independent hourglasses, each running at a different rate. It can be mathematically shown (I will spare you this) that elapsed time can be determined by measuring the sand only in the lower compartments of two hourglasses that run at different rates, so the requirement to measure sand in the upper compartment (by analogy, to measure the uranium isotopes) is eliminated. Therefore, the isotopic analysis of lead in a mineral is all that is required to date its time of formation. In the late 1930s Nier considered using this technique to measure the isotopic composition of lead in a variety of rocks, with the intent of averaging them so as to arrive at a value for the entire crust of the Earth. However, lead occurs in rocks in only tiny concentrations, amounts too small to be determined with the technology of his day. He had to content himself with analyzing galena, the shiny ore of lead. As the Second World War ended, various scientists employed Nier's superb analyses to estimate an age for the Earth of around 3.3 billion years, similar to the upper limit defined by Holmes. This age, although closer to the true value, was still low because it rested on incorrect assumptions, especially about what the isotopic composition of lead was at the time the Earth formed.

The correct decoding of lead's memory was finally made by Clair Patterson, then a graduate student at the University of Chicago. By refining the isotopic analysis technique, Patterson was able to measure the minute quantities of lead that occur in ordinary rocks and minerals, rather than just lead ores. On the suggestion of a professor, he analyzed sulfide (which contained small quantities of lead, but no uranium or thorium) from the Canyon Diablo meteorite, a piece of the object that had excavated Meteor Crater, Arizona. Since the sulfide contained no radioactive source that would produce new lead, its mix of lead isotopes had to have remained unchanged since the meteorite formed. Patterson took this mixture of isotopes as the starting point, the composition of primordial lead in the solar system. With this starting composition, he then used analyses of lead isotopes in sediments deposited on the ocean floor to calculate their age. Because the marine sediments had been eroded and scavenged from vast continental areas, he thought that their composition might represent a global average for the Earth's crust, and thus their age should represent that of the whole planet. Patterson also measured lead isotopes in basaltic lavas from Hawaii, thinking that they might have the same isotopic composition as the mantle from which they were extracted. Both kinds of geologic samples gave ages of 4.5 billion years. Patterson's classic paper, published in 1956, further refined the Earth's age to 4.55 billion years by showing the equivalence of terrestrial lead isotopes with those in meteorites, whose 4.55-billion-year ages had been determined more precisely. The geologic calendar had finally found its beginning.

AN ABSOLUTE CALENDAR

Fixing the time of formation for the Earth merely specifies when the clock started; it does not tell us the chronology for the rest of earthly creation. In fact, placing dates on the remainder of the geologic calendar was continuing all the while the age of the planet was being decoded. In the postwar period, many new radioactive isotopes, including potassium-40 and rubidium-87, began to be applied to rocks. To be a good geologic chronometer, a radioactive isotope has to decay very slowly, over periods of millions or even billions of years. Carbon-14, another isotope that found application as a timepiece, decays much more rapidly, and thus is mostly used to date archaeological artifacts.

Not every rock can be dated, even if it contains radioactive isotopes. The ages of igneous and metamorphic rocks can usually be determined unambiguously, because all the minerals in these rocks formed simultaneously as the rocks crystallized, which effectively set their isotopic clocks running. However, sedimentary rocks are very difficult to date. After all, they formed by assembling fragments of other, preexisting

The Age of the Earth

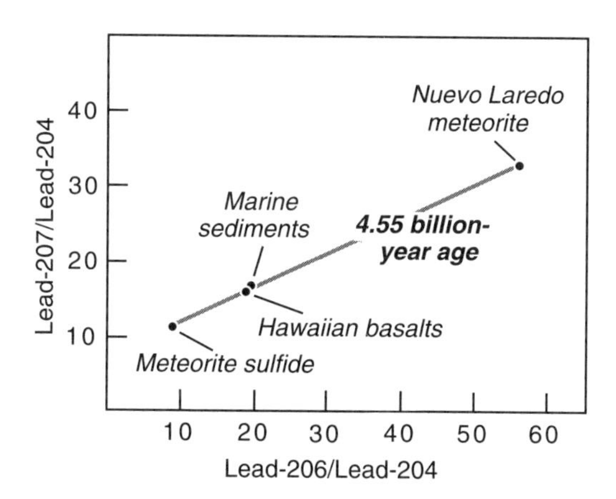

This diagram, adapted from Clair Patterson's classic 1956 paper on the age of the Earth, shows the measured relative abundances of lead isotopes in various materials. Meteorite sulfide, defining one end of the line, contains no uranium so its lead isotopic composition has remained unchanged; it represents the lead isotopic composition at the time the solar system formed. The Nuevo Laredo meteorite, at the other end of the line, has higher proportions of lead-206 and lead-207 as a result of the radioactive decay of uranium. The slope of the line joining the two meteorite samples corresponds to an age of 4.55 billion years. Terrestrial samples, such as crustal sediments and basaltic lavas from the mantle, plotting on this same line (an "isochron") indicate that the Earth shares this same age.

rocks, and each tiny grain from a different source may have a different age. Deposition of these particles and cementing them into a rock do not reset the isotopic clocks of the individual grains. Likewise, fossils are difficult to date because organisms rarely contain radioactive isotopes (they are a health hazard in significant quantities), except for some carbon-14 which is by now long gone. Nevertheless, by using certain geologic tricks, such as bracketing the age of a particular stratigraphic layer by measuring the ages of datable igneous rocks below and above it, geologists have managed to construct a calendar with absolute dates. Once strata containing certain distinctive fossils have been dated (thereby fixing the times when the organisms lived as well), the fossils can be used to assign absolute ages to other rocks that contain them.

Long before geologists had the ability to determine absolute ages for rocks, they constructed a relative geologic time scale, punctuated by important events such as large shifts in global sea level (resulting in major unconformities) or the major extinctions and flowerings of life forms. The basic subdivisions of this scale are called *eras,* and the absolute times

The Geologic Time Scale

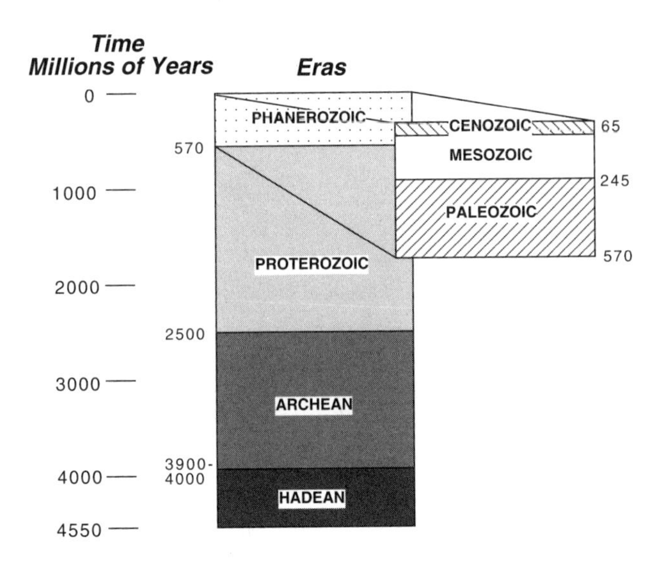

The geologic time scale shows the eras in their relative positions in time, as well as the absolute dates (in millions of years) marking the boundaries between eras. Phanerozoic time contains three eras, shown in the expanded box.

marking their transitions are now well established. The Hadean era represented the earliest stages of our planet, and the planet-forming events we have considered so far took place in this time interval. That was followed, beginning about 3.9 to 4.0 billion years ago when the most ancient known crust formed, by the Archean era. The rest of the eras are named for the kinds of life forms that they hosted. The onset of the Proterozoic (microscopic life) era occurred 2.5 billion years ago. At 570 million years ago this era gave way to Phanerozoic (mesoscopic life) time, which constitutes three eras. The Paleozoic (ancient life) era ushered in the first appearance of organisms with hard skeletal parts. The Paleozoic terminated with a crash of continents and life forms 245 million years ago, marking the transition to the Mesozoic (middle life) era, the time of dinosaurs. The last era, the one in which we live, is the Cenozoic (known life). It too began with a great dying, some 65 million years ago. A significant part of the geologic time scale is represented in the walls of the Grand Canyon, a stratigraphic record already alluded to in the introduction to this chapter.

The eras are further subdivided into *periods,* some of which have familiar names, like the Cambrian or the Jurassic, and the periods into even smaller subunits. There is now little argument about the timing of periods, although the chronology of finer time units is still being determined.

LUNAR ROCKS AND THE AGE OF THE MOON

The Moon probably formed from the debris produced by an impact of a huge projectile with the Earth. Solidification of the lunar crust from a vast magma ocean should have set its isotopic clocks. Unlike our planet, the Moon is geologically dead, and so there is the possibility that some of its earliest rocks might be preserved. The ages of the oldest lunar rocks could serve as a constraint on the timing of the Moon's formation.

A radioactive clock based on the isotopes of two obscure elements (unstable samarium-147 and its decay product, neodymium-143) has been used to determine the ages of crustal samples returned by the *Apollo* astronauts. The principles of this chronometer are similar to those of other radioactive systems such as uranium-lead, so instead of subjecting the reader to the details, I will proceed directly to the bottom line. The oldest reliable ages thus far reported for lunar rocks range from 4.44 to 4.56 billion years, with the older ages having greater analytical uncertainties than the younger ages. A reasonable reading of these data is that the Moon formed soon after the Earth, probably no more than a few tens of millions of years.

Collisional models for the formation of the Moon suggest that the Earth was already differentiated into mantle and core at the time of the impact. Analysis of the isotopes of tungsten, a siderophile element sequestered in the core during differentiation, by Alex Halliday and his coworkers at the University of Michigan suggest that the Earth's core differentiation and the impact that formed the Moon both occurred within eighty million years of the planet's accretion. Planetary differentiation was not instantaneous, but it was reasonably quick. Rapid core formation is exactly what would be expected if heating were mainly caused by impacts. The earliest era of solar system history was obviously a busy and violent time, when the growth and differentiation of planets occurred almost simultaneously.

REFRACTORY INCLUSIONS AND THE AGE OF THE NEBULA

So far we have learned that the Earth is 4.55 billion years old, and the Moon must be only slightly younger. The same 4.55-billion-year birthday is normally ascribed to the entire solar system, although it is probable that there may have been minor differences in the formation times of different planets. But can we really be confident that the solar system itself was not long in the making, that the very construction of the Earth did not require some vast amount of time? It would be helpful to confirm that the Earth is really not much different in age from the materials from which it formed or, stated another way, that the nebula persisted for only a short time.

This has been addressed using several different methods. Most theoretical studies suggest that the collapse of molecular clouds into stars and nebular disks is over and done with rather quickly, generally in less than a few million years. Such calculations do not directly address planet formation, but that is assumed to go hand in hand with disk formation. Once the nebula dissipated, planet accretion, robbed of its raw materials, must have ceased. Astronomical observations of T Tauri stars with disks suggest that some of these stellar babies may be at least 10 million years old, so the time scale for the solar nebula may have been a little longer than theory might suggest.

Another way to assess the lifetime of the nebula is to examine the oldest known materials formed within this environment. There is ample reason to believe that the curious refractory inclusions that occur in chondritic meteorites are the most ancient surviving relics from the nebula. We have already seen how radioactive aluminum-26, which has an extremely short lifetime, was created in a supernova that exploded just before the solar system formed. Aluminum-26 decays into stable magnesium-26, which is rather unfortunate, as magnesium-26 is the common isotopic form of magnesium. Thus the tiny amounts of magnesium-26 created by decay of this short-lived radionuclide get swamped in the much greater amount of magnesium-26 already present in meteorites. Typhoon Lee and his colleagues at the California Institute of Technology seized upon the idea of searching for magnesium-26 inside feldspar, a mineral that contains aluminum but not magnesium. Therefore, any magnesium-26 that is present must have been aluminum, that is, *live* aluminum-26, at the time the feldspar crystallized. The feldspars in refractory inclusions were found to have incorporated more aluminum-26 than any other components of meteorites. With this observation, Lee and his colleagues demonstrated that the refractory inclusions formed earlier than other solar system matter.

The time of formation of refractory inclusions has been determined very precisely by uranium-lead dating to be 4.56 billion years, with an uncertainty of only 0.01 billion. This age is only a few million years older than the age of the Earth. We can conclude, then, that the duration of the nebula must have been short, apparently just a few million years. Thus the meteorite data indicate that the solar nebula formed, collapsed into planets, and dissipated very rapidly.

GALAXIES AND THE AGE OF THE UNIVERSE

The year 1929 is mostly remembered for its stock market crash, but it also marked a discovery of truly astronomical proportions. Edwin Hubble of the Mt. Wilson Observatory in California proved conclusively that the universe is expanding, as galaxies rush away in all directions from other galaxies. As observed from the Earth, the light from all distant galaxies is redshifted, provided that corrections are made for the local motions of stars within galaxies. Moreover, Hubble showed that the amount of this redshift, that is, the relative velocity of each galaxy, is proportional to its distance. This is the signature of an explosion—the exploded fragments of bombs behave the same way. It is no accident that the origin of the universe is described as a "Big Bang." The mathematical factor that relates velocity to distance is now appropriately termed the Hubble constant.

The inverse of the Hubble constant gives the age of the universe, provided that its expansion has occurred at a constant rate. To see why this is so, consider a car traveling directly away from you, so that it is now 100 kilometers away. Using a very powerful radar gun, you determine that the car's speed is 50 kilometers per hour (this is the same principle that astronomers use to gauge a star's velocity). Now let's compute the Hubble constant (defined as velocity divided by distance) for the car. Our determined value is 0.5 kilometers per hour per kilometer, and its inverse is two hours, the time at which the car left the point where we now stand. That calculation is simple enough; arriving at an accurate value for the Hubble constant, however, has proved to be a challenge. Measuring the velocities of nearby stars is almost impossible, because their local motions are so large in comparison to the velocity due to expansion. For faraway objects, however, the local motions become negligible. The real difficulty is in measuring the distances of faraway stars accurately. The most common methods for doing this employ *standard candles,* based on the idea that certain kinds of stars, regardless of their distance, have characteristic intrinsic brightnesses. For example, certain pulsating stars called Cepheids (Polaris, the North Star, is the best-known example) are thought to have high luminosities that are predictable functions of the period in which they expand and contract. Another important method for measuring distance is based on *standard rulers,* objects of constant size whose apparent dimensions decrease with distance. An example of the standard ruler is the spherical region of ionized hydrogen around newborn stars. From these few examples, it should be apparent that there are many ways to estimate the Hubble constant, and the results vary from 50 to 100 kilometers per second per million parsecs (the *parsec* is a common unit of stellar distance, equal to the distance light travels in 3.26 years). When converted to time, these values correspond to ages ranging from 7 to 20 billion years. The most precise measurements of distance, obtained in 1994 from observations of Cepheids using the Hubble Space Telescope (there is something fitting about applying the Hubble Telescope to measure the Hubble constant), give ages ranging from 8 to 12 billion years.

Astronomical observations and theory also form the basis for a second, independent method of assessing the age of the universe. Our own galaxy, the Milky Way, is a gigantic pinwheel containing billions of stars. However, not all of its stars lie within the flattened disk. Scattered around the spiral wheel is a halo of several hundred stellar groups known as globular clusters, each containing hundreds of thousands of stars. Compared to the stars in the galactic disk, the globular cluster stars are deficient in heavy elements, presumably because they formed early, before recycling of heavy elements from massive stars in the disk had increased the average metal content of galactic disk matter. So the globular clusters are thought to represent a sampling of the first stars to form in the Milky Way.

We have already seen how stars evolve, burning first hydrogen (on the main sequence) and then other elements (in the red giant stage), before exploding as supernovas or collapsing into white dwarfs. An important point to be stressed is that massive stars run through their lifetimes much more quickly than do smaller stars. Although less massive stars remain on the main sequence, more massive stars have already peeled off into the red giant field. It is possible to calculate the evolutionary path a star of a given mass would follow on the H-R diagram. If many such paths are calculated for stars of various sizes, curves can be drawn connecting the positions of all the stars of the same age, regardless of their masses. Such curves, illustrated in an accompanying figure, are called *isochrones*. Superimposing isochrones onto an H-R diagram for stars in a globular cluster allows us to estimate the age of that cluster. The point on the diagram at which stars peel off the main sequence to reach into the red giant field is particularly diagnostic. Based on this technique, globular cluster M92 appears to be about 18 billion years old. Other stellar clusters in the Milky Way give ages ranging from 14 to 18 billion years. Although this time strictly dates the time of formation of our galaxy, rather than the universe, all the galaxies are thought to have formed at about the same time. The mismatch between these ages and the most recent age estimates based on the Hubble constant (that is, the apparent age of our galaxy is greater than that of the universe?) has caused some consternation among astrophysicists.

Finally, a third method of dating the universe is based on the ages of certain very long-lived radioactive isotopes. Let us consider, for example, the value of the ratio of thorium-232 to uranium-238, which at the present time equals four. Because the rate at which each isotope decays is different, this ratio has changed over time. Furthermore, we know the rate of decay, so we can calculate the ratio at any time. This calculation is complicated, though, by the fact that neither isotope was produced in the Big Bang at the beginning of the universe; instead, both have been created continuously during supernova explosions. The age of the galaxy determined in this way is constrained to lie between 13 and 20 billion years.

Taking all these astrophysical and isotopic results together, we obtain a spectrum of ages ranging from about 8 to 20 billion years. Many astrophysicists seem comfortable with an age toward the middle of this range (most methods overlap at 12 to 15 billion years), but the issue is not decided.

AND THE REST IS HISTORY

In explaining the rules for the game of science in 1868, Thomas Huxley enunciated some cold, hard facts for those who would uncover scientific truths:

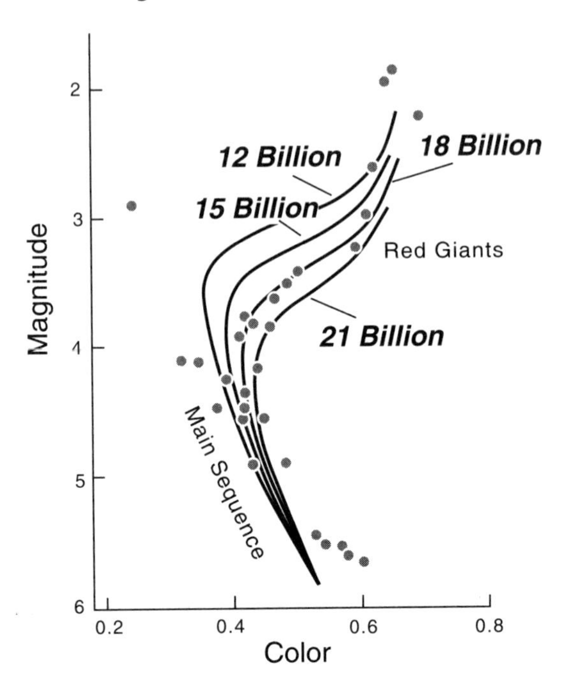

Dots in this H-R diagram show the luminosities and colors of stars in the M92 globular cluster, a patch of the earliest-formed stars in the Milky Way. The superimposed curves represent calculated isochrones, that is, lines along which all stars of the same age should plot. Ages of the isochrones are labeled in billions of years. The point at which the stars in this cluster turn off the main sequence corresponds to an age of about 18 billion years. This diagram uses slightly different axes from the H-R diagram in a previous chapter: its horizontal scale is color index (the ratio of blue to yellow light received from stars) and the vertical axis is magnitude (a scale of relative brightness). (Modified from A. Sandage [1983], *Astronomical Journal*, vol. 88)

> The chessboard is the world, the pieces are the phenomena of the universe, the rules of the game are what we call the laws of nature. The player on the other side is hidden from us. We know that his play is always fair, just, and patient. But also we know, to our cost, that he never overlooks a mistake, or makes the smallest allowance for ignorance.
>
> —T. H. Huxley, *A Liberal Education*

This quotation could serve as a primer for the study of the chronology of creation. There have been many mistakes and much ignorance displayed in unraveling the laws of nature that apply to measuring time, and progress has been slow and hard won. However, thanks to the efforts, advances, and mistakes of many, the history of creation can now be written in absolute numbers, though admittedly with large error bars in some cases.

The universe was created perhaps 12 to 15 billion years ago. The Milky Way galaxy formed soon thereafter. Within this arena of stars, a molecular cloud collapsed approximately 4.56 billion years ago, forming the solar nebula. A supernova exploded in the neighborhood, possibly precipitating this collapse, and its short-lived radioactive ashes were incorporated, still alive, into the earliest condensed nebular solids. The gaseous and solid matter comprising the nebula was rapidly swept up into planetesimals, which in turn accreted into larger planets. Within a span of a few million years, the dusty nebula was gone, replaced by a gleaming stellar centerpiece and a retinue of planets. The age of one of these, the Earth, has been precisely determined at 4.55 billion years. Rapid heating during accretion caused the Earth to differentiate into rocky mantle, crust, and metallic core. Shortly thereafter, the planet was struck by a huge projectile. A large Moon rapidly accreted from the vaporized rubble of this collision, so that it was intact by at least 4.50 billion years ago. Thus ended creation, and began a much slower evolutionary transformation of the Earth into the world we recognize today.

Some Suggested Readings

Allegre, C. J. (1992). *From Stone to Star: A View of Modern Geology*. Harvard University Press, Cambridge, MA. *Chapter 3 of this book gives a lucid account of the construction of the geologic time scale.*

Allegre, C. J., Manhes, Gerard, and Gopel, Christa (1995). The age of the Earth. *Geochimica et Cosmochimica Acta,* vol. 59, pp. 1445–56. *This technical paper gives the most up-to-date summary of precise age-dating applicable to the Earth; great science, but not for beginners.*

Burchfield, J. D. (1975). *Lord Kelvin and the Age of the Earth.* Science History Publications, New York. *A fascinating history of Lord Kelvin's ideas and influence in shaping our understanding of geologic time.*

Dalrymple, G. B. (1991). *The Age of the Earth.* Stanford University Press, Stanford, CA. *This authoritative treatment describes in detail the evolution of thought on the age of the Earth; especially appropriate are Chapter 1 on the early attempts to assess geologic time, Chapter 7 on the use of lead isotopes as a chronometer, and Chapter 8 on the age of the universe.*

Dickin, A. P. (1995). *Radiogenic Isotope Geology*. Cambridge University Press, New York. *A comprehensive textbook explaining the principles of most modern chronometers based on radioactive isotopes.*

Podosek, F. A., and Cassen, Patrick (1994). Theoretical, observational, and isotopic estimates of the lifetime of the solar nebula. *Meteoritics*, vol. 29, pp. 6–25. *This superb technical review of constraints on the age of the nebula covers a lot of ground and presumes some technical knowledge.*

Terra Not Firma

Origin and Evolution of the Continents

FOREIGNERS IN THE NEIGHBORHOOD

On a clear day, looking south from any high vantage point in my hometown of Knoxville, one can easily spy the hazy ridges of the Great Smoky Mountains. Even from a distance, these towering Appalachian crags almost seem to leap onto the undulating foreground. And, once upon a time, that is precisely what they did. The western edge of the Smokies is actually a great thrust fault, along which the rocks comprising the mountains have been pushed westward for hundreds of kilometers over the foreland. In a few places, like picturesque Cades Cove in the Great Smoky Mountains National Park, erosion has cut a window through the overlying rocks, exposing the foreland underneath. This incredible crumpling of foreign rock, nudged almost into my backyard, was shoved to the west when what is now Africa slammed into North America 265 million years ago.

Much farther to the east, in the Carolinas, lies the actual suture where the two continents were welded together for a time, as well as down-dropped blocks and upwellings of basaltic magma that formed when they again split apart some 50 million years later. Since that time, North America and Africa have continued to migrate slowly away from each other, over time creating the ever-enlarging gulf we call the Atlantic Ocean. Other continents, too, have shifted about the planet's surface almost at random, sometimes rattling around as relatively small pieces but occasionally amalgamating into huge land masses.

At present, the continental crust (including the submerged continental shelves) comprises only 41 percent of the Earth's surface area and an almost trivial 0.35 percent of its total mass. However, this material is particularly important for reconstructing the planet's history, not only because continents are easily accessible for geologic study but also because their great antiquity relative to the oceanic crust means that they carry the bulk of the geologic record. In this chapter we will examine how and when continental crust was first created, as well as how it has grown in volume and changed its form over geologic time. We will also see that the continents are made of rocks with very peculiar compositions, and we will explore how they came to be this way. It is important to remember that the continents are the abode of many highly evolved life forms, and without these shifting platforms of dry land the living world would be unrecognizable.

AGROUND OR ADRIFT?

The famous children's story *The Little Prince* describes a boy who visits other planets. On the sixth planet he encounters an old man writing books, a geographer who asks the visitor to describe his tiny, asteroidal home.

> "I have three volcanoes. . . ." [the boy replied.] "I have also a flower."
>
> "We do not record flowers," said the geographer.
>
> "Why is that? The flower is the most beautiful thing on my planet!"
>
> "We do not record them," said the geographer, "because they are ephemeral."
>
> "What does that mean—'ephemeral'?"
>
> "Geographies," said the geographer, "are the books which, of all books, are most concerned with matters of consequence. They never become old-fashioned. It is very rarely that a mountain changes its position. It is very rarely that an ocean empties itself of its waters. We write of eternal things."
>
> —Antoine de Saint-Exupéry, *The Little Prince*

Until the middle of the twentieth century geologists, as well as geographers, regarded the Earth as rigid and the continents as fixed. In the early part of this century, Alfred Wegener had a contrary notion.

Wegener was actually trained as an astronomer, but while a student he became fascinated with the new science of meteorology, which he subsequently practiced at universities in Germany and Austria. He was also an explorer and an adventurer, participating in several Arctic expeditions to the Greenland ice cap and setting a world record for a balloon flight. Maps are part of the stock and trade of explorers, and Wegener spent considerable time poring over their details. He was struck, as were many others ahead of him, by the similarity of opposing coasts abutting the Atlantic Ocean. What made Wegener different was that he offered a bold conjecture about the meaning of this puzzlelike fit. In *The Origin of Continents and Oceans,* published in German in 1912 and translated into English in 1915, he argued that a great continent had been rent and had drifted apart to form the Atlantic basin.

This idea was heresy to geologists, as it would have been to the geographer in *The Little Prince.* Current thinking envisioned a world still cooling and solidifying and, in the process, shrinking, so that the crust was warped up and down by the forces of contraction, much like the skin of a dried apple. Arching of the crust in some places caused its collapse at other locations and, in so doing, the ocean basins were thought to have been created. Since vertical crustal movements were manifest everywhere, geology was enamored of the idea of vanished land bridges to explain the faunal connections between separated continents. But there was no room in geology for horizontal crust movements; the continents were thought to be forever fixed in outline and position.

Wegener bolstered his controversial claim of meandering continents by amassing an impressive body of supporting evidence. Being a meteorologist, he was particularly swayed by indications of past climate change. He argued that in places as widely separated as South America, Australia, India, Africa, and Antarctica, deposits of rubble from receding glaciers indicated that all these continents had experienced polar climates at about the same time. On other continents now at high latitudes, he cited the occurrence of salt deposits, formed by evaporation in arid regions, as proof of their drift from equatorial latitudes.

This evidence received a sympathetic hearing in Europe. On the other side of the Atlantic, however, the idea of drifting continents was ridiculed. Wegener was put on the defensive at a symposium, held in the United States in 1926, where virtually everyone rudely attacked his ideas. To cite but one example, Thomas Chamberlain, an enlightened champion of an ancient age for the Earth, stated, "If we are to believe Wegener's hypothesis, we must forget everything that has been learned

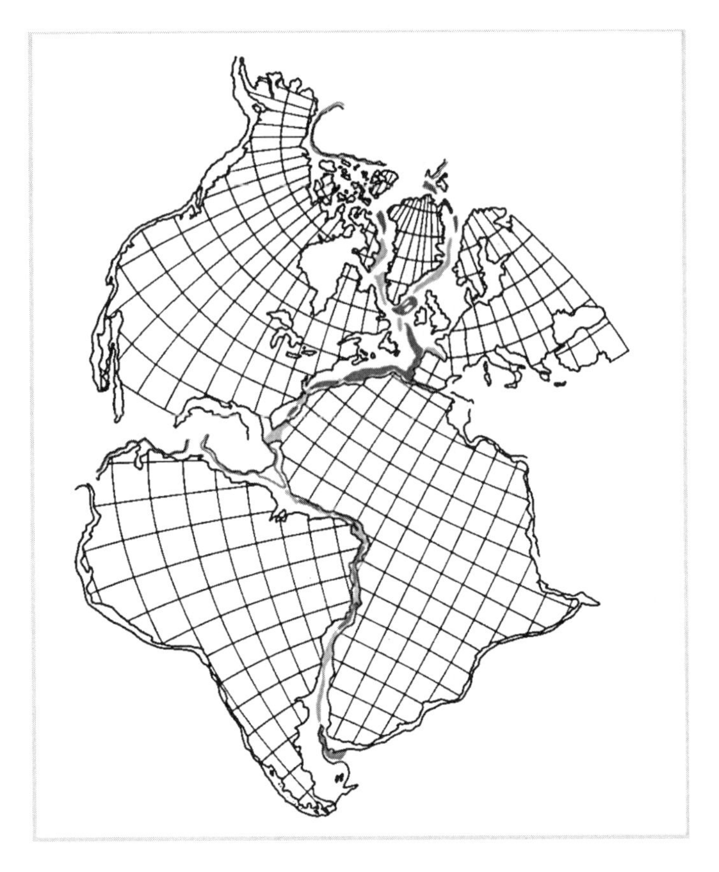

The remarkable geometric fit of continents on opposite sides of the Atlantic convinced Alfred Wegener that they were once connected. In 1965 several British scientists constructed this rigorous match by including the offshore shelves as parts of the continents. Solid areas indicate where the continents do not quite fit together. (After E. C. Bullard, J. E. Everett, and A. G. Smith [1965], *Philosophical Transactions of the Royal Society of London,* vol. A258)

in the past seventy years and start all over again." Having just been reconciled to the abyss of geologic time, science was not yet ready to deal with drifting geography.

Wegener was especially called to task on the weakest link in his argument, the mechanism by which the continents moved. He envisioned land masses as somehow plowing through the oceanic crust, as if they were stony icebergs in a sea of weaker rock. The force that propelled them foward, and also that compressed the crust into great mountain chains, Wegener called *Polflucht,* or "flight from the poles." He suggested that continents raised above sea level on a spinning Earth should move toward the equator and to the west. Tidal forces acting in this direction do exist, but it was almost immediately recognized that they are far too puny to deform rocks. Physicist Sir Harold Jeffries of the University of Cambridge correctly pointed out that the ocean floors were much too rigid to allow for the passage of moving continental masses. Without a plausible physical means for moving continents about, the puzzlelike fit of the continents, however striking, was considered fortuitous.

Wegener's proposal languished, relegated to the museum of quaint and outrageous scientific hypotheses. He was clearly too far ahead of his time, but it did not help that he had no credentials as a geologist and so was considered something of an amateur by those whose view of the world he had challenged. Not until about 1960 did geologists finally become so overwhelmed with evidence that they had to take a more serious look at continental drift.

The first hint that something was amiss with the notion of rock-solid continents came from studies of rock magnetization. As magmas crystallize, iron-bearing mineral grains in them, just like tiny compasses, record the orientation of the Earth's magnetic field at the time. The direction of the ancient magnetic field can be measured in rocks if their orientations are recorded before they are prized loose from an outcrop. When measurements are made on at least three rocks of the same age in different locations, the position of the magnetic pole at that time can be fixed by triangulation. The inclined angle that the rock's magnetic vector makes with the horizontal also gives information about its latitude at the time it solidified. Soon after such data began to be collected, some curious relationships emerged. When the magnetic pole positions determined from rocks of different ages on the same continent were plotted, they did not coincide with the present pole position. Instead, they traced out a curve on the globe, as if the magnetic pole were wandering. Even more perplexing was the fact that rocks on different continents appeared to define different polar wandering curves, separate paths that finally converged at the present pole position. It seemed ludicrous that each continent would have its own magnetic pole. However, if the continents were assembled together millions of years ago and then had drifted apart, as Wegener suggested, all the polar paths could be made to collapse into a point at the present pole position. The magnetic pole only appeared to wander, when in fact it was the continents that moved.

In 1960 Harry Hess, a geology professor at Princeton University, proposed a plausible mechanism by which the seafloor could spread. This novel idea, at the same time, allowed for continental drift. Hess envisioned that internal heating caused upwelling of solid mantle material, a heat transference process termed convection. A boiling pot of soup convects when hot liquid from the bottom of the pan ascends and breaks the surface, and cooler soup sinks along the sides. Hess argued that when the convecting mantle approached the Earth's surface, it then moved laterally, and after some distance finally cooled and descended again into the mantle. He also suggested that the oceanic crust might ride on the flowing mantle like a raft. In the gap where parts of the crust pulled apart (called a *spreading center*), lava would erupt and form a mid-ocean ridge. At the other end of this conveyor belt, excess oceanic crust must be dragged down into the mantle (a *subduction zone*).

Almost simultaneously, geophysicists provided proof of Hess's idea, again in the form of rock magnetism. For years research ships had towed magnetometers behind them, recording the magnetization of the seafloor. The Earth periodically reverses its magnetic field, north to south and back again, and these reversals stand out clearly in volcanic rocks on the ocean bottom. Maps of the rock magnetic patterns, if colored white or black depending on the magnetic polarity, look like a zebra skin, with symmetrical patterns developed on each side of the mid-ocean ridge axis. The obvious explanation for this striking pattern, first provided by Drummond Matthews and Fred Vine of Cambridge University, was that the ridge axis on a spreading seafloor was filled by basaltic magma that recorded the orientation of the magnetic field at the time of its eruption. When the congealed rock later split apart, the separating plates carried symmetric halves of the magnetic anomaly, and new magma was inserted between them, now recording a change in magnetic polarity.

The identification of locations where plates are recycled into the mantle completed the story. The subducted plates are cold and brittle, and so they fracture, producing earthquakes. By pinpointing where the earthquakes originated, it is possible to image the subducted plate. Faced with this burgeoning evidence, the geologic community embraced the concept of *plate tectonics*. But that is perhaps too mild a description for this revolution; at about the same time that the Beatles turned popular music on its head in the early 1960s, geology found itself in the same position that astronomy was in the time of Copernicus.

Plate tectonics was critically important for the subject at hand, that is, drifting continents, because it provided a way to move them about. The continents are actually embedded within the moving plates, and so move along passively as the seafloors spread. There is no need to argue, as Wegener did, that continents plow through stationary oceanic crust.

Plate Tectonics

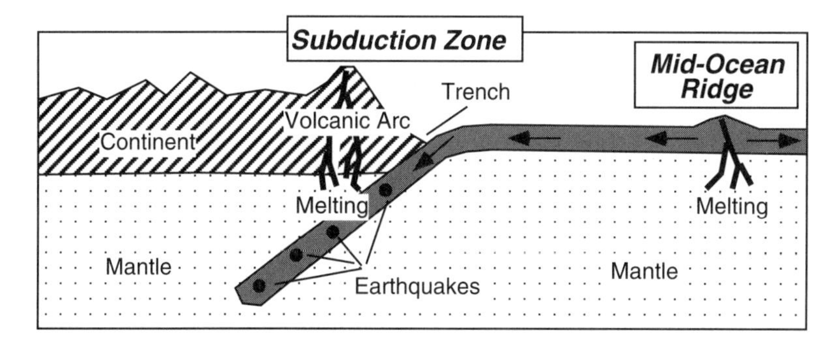

This cross section of the crust and upper mantle shows a mid-ocean ridge, where new oceanic crust is born, and a subduction zone, in which the moving plate is consumed in the mantle. The subducted slab can be imaged by plotting the locations of earthquakes. Melting in the subduction zone produces volcanoes on the overlying plate (a continent in this illustration), thereby creating more continental crust.

At the other end of the conveyor belt, continents either ride above descending slabs of denser oceanic crust or are driven into the edges of other continents. The latter process is responsible for assembling large land masses, just as North America and Africa were once welded together in my backyard.

CONTINENTS RESHUFFLED, THEN AND NOW

At one time or another, most people have peered through a kaleidoscope. The changing geometric patterns are mostly a trick of the mirrored reflections of colorful glass shards, but they also depend on the random movements of these fragments as the tube is twisted. Different shards sometimes touch, often slide past one another, and at other times lie at opposite sides of the kaleidoscope.

That is not a bad analogy for the nomadic behavior of continental crust over geologic time. As one well-studied example, let's examine Pangaea (the "all-land"), the supercontinent first envisioned by Wegener. Toward the end of the Paleozoic era, about 265 million years ago, most of the world's continental crust was assembled into a huge, collective mass. The largest constituent piece, called Gondwanaland, consisted of most of the continents (and one subcontinent) in the present Southern

Trace of the South Pole Across Gondwana

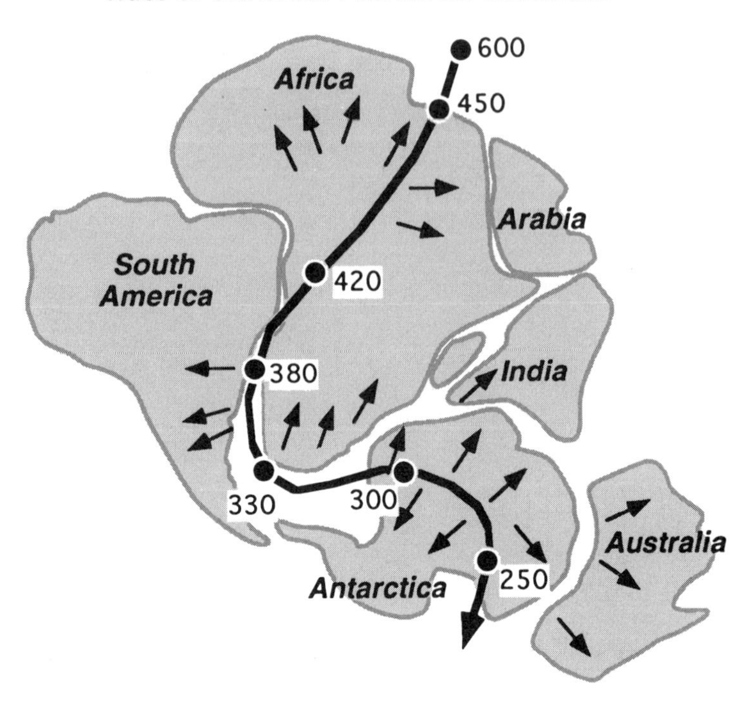

Gondwana was assembled from the present continents and subcontinents of South America, Africa, Antarctica, Australia, India, and Arabia. This landmass traversed the South Pole during the Paleozoic era; its progress is indicated by the times (in millions of years ago) along the track. Small arrows mark the flow directions of glaciers that covered parts of the supercontinent at different times.

Hemisphere—South America, Africa, Antarctica, India, and Australia—that had been together for at least several hundred million years.

Gondwana itself moved around over the face of the globe during the latter part of the Paleozoic era, as indicated by the wandering trace of the South Pole (actually the magnetic pole, which corresponds approximately to the geographic pole). As the part of Gondwana that we now recognize as Saharan Africa marched over the pole about 440 million years ago, glaciers scoured the surface, leaving telltale traces that ice sheets once ruled the desert. The icecap simultaneously covered parts of southern Africa, Brazil, and Antarctica about 330 million years ago, and finally vanished 70 million years later as the edge of Antarctica cleared the pole. These "ice ages," formerly thought to represent dramatic shifts in the global climate, actually represent periods when parts of Gondwanaland squatted over the South Pole.

As Gondwana moved northward, it encountered other large land masses, Laurentia (now North America and Greenland) and Baltica (now parts of Eurasia). These, too, were added to the growing supercontinent, so that at its peak it stretched almost from pole to pole. At the same time that Antarctica, on one end, was covered with ice, more equatorial parts of the land mass experienced harsh, arid conditions. Inland seas evaporated to form salt deposits, and dunes of sand shifted across desiccated landscapes.

Continents join because they happen to collide, as they ride passively along on moving plates. Why they break apart is less obvious, but just as dramatic. At the beginning of the Mesozoic era, Pangaea began to self-destruct, after having existed in its full glory for only about 50 million years. Fractures appeared in various places as the giant continent overrode the sites of mantle plumes. There the crust was domed upward and, as it arched, it cracked in several directions. Some of the cracks gradually widened into rifts, where the crust stretched so far apart that central blocks collapsed. Sometimes upwelling basaltic magma erupted into the rifts. Eventually the rifts from separate domes connected with those of neighboring domes, forming extensive breaks that caused the supercontinent to shatter like a glass windowpane. The largest breaks ultimately filled with water and became oceanic spreading centers, such as the Mid-Atlantic Ridge. Different rift segments, of course, opened at different times, so that the South Atlantic yawned apart later than the North Atlantic.

Pangaea was not the first supercontinent, nor will it be the last. Many of the world's continents were combined into a more ancient continental giant, called Rodinia, about 750 million years ago. But amalgamation is not restricted just to the grand scale of supercontinents. Even the dispersed continents of today's globe were assembled piecemeal from smaller pieces at one time or another. Geologic mapping of the continents shows that they usually contain one or more *cratons,* which are very old (Archean era) continental nuclei, to the edges of which have been attached terranes of younger rocks.

Sometimes we see many such terranes, plastered layer after layer onto a continent's outer margin. Westernmost Canada and Alaska have been constructed from wave after wave of accreted terranes, each with its own distinctive geologic history. This suggests that crustal flotsam and jetsam eventually must be gathered in subduction zones, like floating scum in stream eddies, to be welded onto whatever continent may be lurking nearby. The successive times when each terrane docked to the

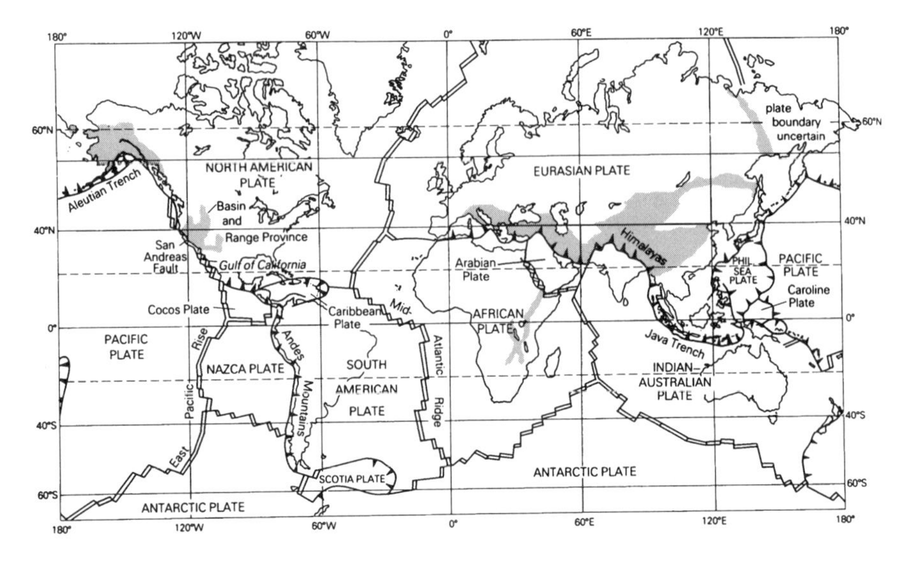

The Earth's surface is presently broken into a dozen large plates, bounded by spreading centers (double lines), subduction zones (lines with teeth pointing in the direction of subduction), and transform faults where plates merely slip past each other (single lines). (U.S. Geological Survey)

mainland can be defined by painstakingly studying the ages of sediments that overlap onto the adjacent terranes.

The structures of continents, with ancient cratons nested inside belts of rock accreted at later times, suggest that plate tectonics is not a new phenomenon. The exact time at which the Earth's exterior shell first cracked and the broken slabs began to move about and jostle each other is a subject of some controversy, but it seems safe to infer that the continents have been on the move for much of our planet's history.

At the present time the Earth's surface is divided into twelve large plates. Viewed from on high, we can now actually measure their relative movements. Surveying techniques have advanced to a point where it is possible to detect differences of a few centimeters in the distance of two points that are thousands of kilometers apart. Using a global satellite positioning system, the rate at which the Atlantic is spreading has been determined to be a few millimeters per year. Such measurements also carry some surprises. For example, it is now known that the rate at which a plate moves varies inversely with the proportion of continental area on the plate. This may relate to the fact that continents are underlain by thick mantle keels that remain forever attached. Because larger continents have larger roots, it is thought that the keels must exert a drag, slowing the motions of the plates that house them.

TAKEN FOR GRANITE

Although plate tectonics continues, to this day, to redistribute continents about the globe, it does not provide an obvious explanation for how this crust was created in the first place. A key to understanding how continental crust formed is its composition.

The colorful intricacy of geologic surface maps immediately reveals that continents are highly complex. Their horizontal complexities result, in part, from the piecemeal way they have been put together, as new crust is welded onto the edges of ancient cratons. But there is vertical complexity, too.

The average thickness of the continental crust is about 36 kilometers, at least quadruple that of the oceanic crust. However, continental crust may be as thin as 10 kilometers in some places and as thick as 80 kilometers in others, depending on the size of the continental block and its age. The base of the crust is usually defined as the Moho, the seismic discontinuity where P-wave velocities jump to about 8 kilometers per second. In some areas, though, the Moho is a gradual transition or may even be absent. The upper crust can sometimes be delineated from the lower crust by a mid-crustal boundary, at a depth of 10 to 20 kilometers, called the Conrad discontinuity.

Despite its complexity, the chemical composition of the upper crust is reasonably well characterized. More than half a century ago, geochemist Victor Goldschmidt proposed that fine-grained sedimentary rocks, like shales, should have sampled vast areas of the crust, and so should contain uniform, average crustal abundances of most elements. He noted that water in rivers and in the sea contains few dissolved elements in any significant amount, concluding that most elements are transported and deposited as sediments without loss. Goldschmidt's prediction has proved to be correct, and the upper crustal abundances of most elements are now well established from analyses of sediments. The composition of the lower crust is more difficult to determine, because it is only rarely exposed at the surface.

Although the continental crust contains many different kinds of rock, its average composition is often described as "granitic." Granites are igneous rocks rich in *silica* (silicon dioxide—the mineral quartz in pure form, but also combined with many other elements to make other silicates). The crust, like granite, also has high concentrations of aluminum, sodium, and potassium, all important components of feldspars, the most common silicate mineral in the crust. As minerals go, feldspars and quartz have relatively low densities, so the average density of the continental crust is significantly lower than that of the basaltic oceanic crust. Fragments of lower crustal rock that are carried to the surface by volcanic eruptions suggest that this region contains a higher proportion of basalt than does the upper crust, so its composition may be intermediate between those of the ocean floor and the upper continental crust.

Mother Nature shows a remarkable ability to concentrate certain elements. Fish, for example, can concentrate mercury to unhealthy levels, a discovery that frightened many seafood consumers a few years ago. Of course, the concentration of elements has its good points, too. We are dependent on geologic processes to sequester valuable metals into mineable ore deposits and organic matter into productive fields of fossil fuels. The strategies that nature uses to concentrate elements are diverse. The high silica abundance of the continental crust results from the segregation of this constituent during partial melting, as a melt usually has a higher silica content than the unmelted residue.

Certain relatively rare elements, whose geochemical behavior is described as *incompatible* (meaning that, because of their size or charge, they do not fit comfortably into the crystal structures of most minerals), exhibit truly remarkable enrichments in the continental crust. Such elements in crustal rocks may be enriched by factors of several hundred over typical mantle rock values. When rocks are partially melted, the incompatible elements they contain are sequestered in the liquid, leaving the solid residue depleted. The concentrations of incompatible elements are so extreme that we might regard the continents as a sort of purged scum that must have been partially melted not just once, but repeatedly, in order to produce such extraordinary abundances. Potassium, uranium, and thorium, elements whose radioactive isotopes account for much of the Earth's heat production, are incompatible, and the crust may contain a quarter to a half of the entire planet's inventory of these elements. Although many incompatible elements occur in only trace quantities, even in crustal rocks, their concentrations in sediments (and thus in the average crust) can tell us much about how the crust itself formed.

The mantle keels below the continents differ markedly from the mantle under the oceans. Study of samples of these mantle roots brought to the surface in volcanic eruptions indicates that they are residues left from partial melting. However, the amount of magma extracted from the keels was much greater than that in the mantle source for oceanic crust. It thus seems likely that at least the lower continental crust may have formed by melting the mantle immediately below. The repeated melting events that seem to be required to explain the high abundances of incompatible elements suggest that the upper crust may have formed by remelting of the lower crust.

AGE OF THE CONTINENTS

My mother guards her age like it was a state secret. We sometimes tease her about it, but she is adamant that her birthdate is nobody's business. Continents, too, resist most attempts to pry out the chronology of their birth.

As determined from measurements of radioactivity in the few samples of ancient crustal rocks that seem willing to reveal their true ages, the continents span almost 90 percent of the time that the Earth has been in existence. Portions of cratons are documented to be as old as 3.5 to 3.8 billion years, and tiny zircon crystals in one sandstone from Australia give ages of 4.3 billion years. The sandstone itself, however, is not this old; the zircon grains it contains were weathered out of older igneous rocks and incorporated into this later sediment.

Most of the radioactive chronometers in ancient rocks have been reset by later heating or melting events. The Earth is, after all, a geologically active world, and it is unusual for rocks to have survived for billions of years without experiencing some kind of later thermal processing. As a consequence, the ages determined for specific rocks offer poor testimony to the actual timing of continent formation. A better, though more indirect, approach to understanding the ages of continents comes from studying the sediments derived from them. The specific clock used in this case is the change in the isotopic composition of neodymium over time, as a radiogenic isotope is added to it from the decay of radioactive samarium. Both neodymium and samarium are incompatible elements, and so the neodymium isotopes in sediments relate to the time in which they were first added to the continental crust as magmas, rather than to some later time when the sediment was deposited. Of course, a sediment is a grand average of many such continent-forming events, but from it we can determine the *crustal residence age,* which is how long the neodymium has actually been part of the continental crust.

The concept of residence age is a difficult one, and an analogy may help in understanding this. Let us say we are trying to determine the time at which a old graveyard was first established, as well as the history of its subsequent use. The date of death on the oldest gravestone defines, within a few days, the first burial in the cemetery. From the dates on all the gravestones, we could determine how long each tomb had been in use, and by averaging all those times, the mean cemetery residence age. The residence age represents the average time elapsed since all remains in the burial ground were interred, which has some value intermediate between the oldest and the most recent burials. Similarly, we could calculate the residence age at any time in the past by averaging the burial dates of only the gravestones existing at that time, and so indirectly see how quickly or slowly the cemetery grew.

Now let's return to the more scientifically interesting (and less morbid) situation of crustal residence ages. The neodymium isotopic composition of sediments serves as a time marker, as does the date of death on a gravestone, that can be used to determine the crustal residence age for this element. We will examine how the crustal residence age varies as we sample and analyze sediments that were deposited at different periods in Earth history. A compilation of such measurements shows that the most ancient sediments deposited 3.5 to 3.8 billion years ago have crustal residence ages of about that same value, implying that very little continental crust must have existed before that time. That makes sense, as judged from the impact record of the Moon, because intense meteor bombardment of the Earth-Moon system prior to about 3.8 billion years ago would probably have destroyed most of the crust produced. In contrast, sediments deposited in more recent times have progressively younger crustal residence ages, although the residence times remain older than the times of their deposition. For sediments forming right now in modern environments, the calculated crustal residence age is about 1.7 billion years.

From this pattern we can infer that significant amounts of continental crust first appeared in the neighborhood of 3.8 billion years ago (although ancient Australian zircons suggest some minor crust existed prior to that time). The old crustal residence ages of more recently deposited sediments, including those that are forming now, mean that older continental crust has been repeatedly reworked and mixed with newer additions to the crust. If that were not so, the crustal residence age would never change, regardless of when the sediment formed. Therefore, the amount of continental crust must have progressively increased over time. If the crustal residence age of sediments deposited at any given time in the past is taken as the mean age of the continents at that time, then a curve describing the growth of continental crust can be constructed. The result suggests that growth was faster during the first several billion

Ages of Crustal Sediments

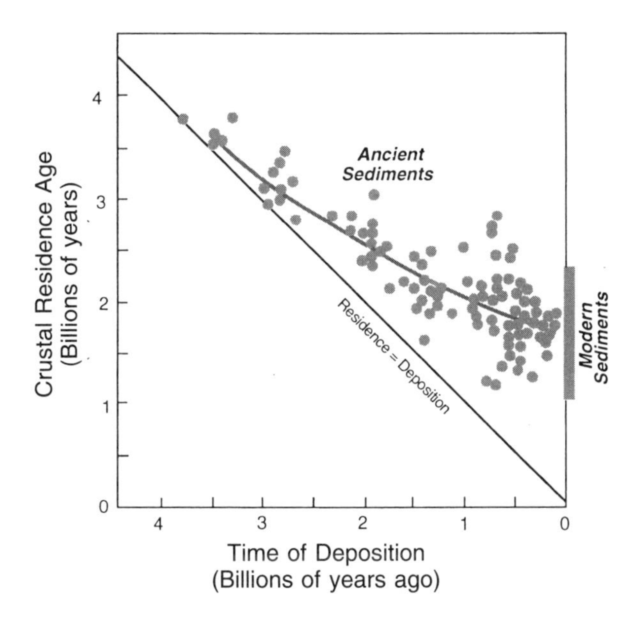

This diagram compares the crustal residence ages (determined from neodymium isotopes) of sediments with the times of their deposition. The diagonal line illustrates where these two values are equal. Modern sediments, shown by the gray bar at the right, have crustal residence ages of approximately 1.7 billion years, implying that they are mixtures of new crust with ancient, recycled crust. Sediments deposited in the geologic past generally have even older crustal residence ages. The most ancient sediments have crustal residence ages that equal the time of their deposition, suggesting that there was little crust before about 3.8 billion years ago. (Modified from K. O'Nions [1992], *Understanding the Earth,* Cambridge University Press)

years, when perhaps 60 percent of the crust formed, than in the succeeding interval.

Recent radiometric dating of rocks from the mantle keels below continents shows that they were melted at about the same time that the oldest (Archean) rocks in the overlying continent formed. This supports the suggestion that the Archean crust formed directly from partial melts of the underlying mantle, and the residues have remained firmly attached to the continents as keels since those events.

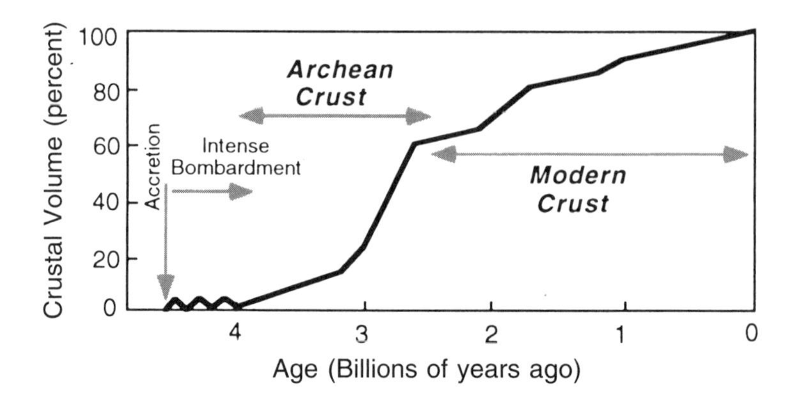

Crustal residence ages indicate that the volume of continental crust has increased over time. About 60 percent of the crust was created during the Archean era, by direct melting of the mantle. The remainder of modern crust formed later, by remelting in subduction zones. Steeper parts of the line represent major episodes of crustal growth. (Modified from S. R. Taylor and S. McLennan [1996], *Scientific American,* vol. 274)

THE OLDEST CRUST ON EARTH

Deep in northwestern Canada lies the Slave craton, one of several Archean terranes nestled in the interior of the North American continent. This craton is composed partly of granite and partly of "greenstone," the latter a metamorphic rock that was formerly basaltic lava. In this regard the Slave craton is not unusual; most Archean terranes contain similar rocks. The granite and greenstone intruded into or have erupted over older rocks, called gneisses. Most gneisses are banded and highly deformed rocks with complex geologic histories, but they could have formed from even older granites and greenstones.

Some of these gneisses are exposed along the shore of the Acasta River, and so they have been given the name Acasta gneisses. These rocks were once igneous, but they have been metamorphosed almost beyond recognition. They are the oldest known rocks on Earth.

Metamorphism resets the radioactive chronometers in most rocks, so determining the time of crystallization of the original igneous rocks requires the analysis of some mineral that survived the metamorphism unchanged. In this particular instance, that mineral is zircon. Lead isotopes in zircons from the Acasta gneiss give ages of 3.96 billion years. This measurement required highly precise analyses of the centers of in-

This satellite photograph of the Pilbara region of northwestern Australia shows a typical craton, with rocks as old as 3.7 billion years. The large rounded blobs are granites, which intruded the surrounding greenstone belts. (NASA)

dividual crystals that are no larger than a pencil point. That was accomplished in 1989 by Samuel Bowring, then a geology professor at Washington University in St. Louis, using an incredibly sensitive instrument at the Australian National University.

It would be easy to read too much into this one small exposure of Archean continental crust, but many other, only slightly younger Archean terranes elsewhere in the world are composed of rocks with similar compositions. Moreover, the incompatible elements in Archean sediments support the idea that the earliest continents formed from two different kinds of rocks. It seems fair to conclude that the earliest continental crust was constructed from both basaltic and granitic magmas. The basalts are thought to have formed by melting mantle rocks in the continental keels, and the granites from melting basalts. The important point is that the earliest continental crust was relatively primitive material, in effect first-generation magmas.

Many geologists believe that Archean crust formed in ancient subduction zones. The granite that characterizes these regions is similar in composition to modern granitic magmas formed from melting subducted slabs that have been altered by seawater.

CRUST NOW IN THE MAKING

The volcanic Andes mountains, extending from Central America down the western flank of South America to its very tip, are places where copious amounts of granitic magma erupt or are intruded into deeper levels of the crust. This is one of many locations where new continental crust is forming at the present day.

Deep under the Andes, oceanic crust is being subducted at a rate of a few centimeters per year. The consumption of this plate crumples the overlying crust into folded mountains, the eastern ranges of the Andes. At a depth of about 150 kilometers, the descending slab is metamorphosed, in the process losing some of the water it contains. The water rises into the overlying mantle wedge, and the addition of fluids causes melting by lowering the melting point of mantle rocks. The resulting magmas are thought to have basaltic compositions. Being less dense than the surrounding mantle, they form huge, inverted teardrops that migrate toward the surface. However, basaltic magma is more dense than granitic crust, so it becomes stalled and ponds at the bottom of the overlying continent. This affords an opportunity for magma to mix with the solid crust, forming a hybrid composition. Crystallization of the basaltic magma also releases heat, which causes the adjacent lower crust to melt. This results in granitic magmas that rise and erupt onto the surface, creating the volcanoes of the western Andean cordillera. As awe-inspiring as these great volcanic edifices are, they are deceiving. Only a tiny portion of the magma added to the crust actually makes it to the surface, so the erupted materials represent only a small fraction of the total increase in crustal volume.

Over time, the volcanoes are eroded, and the sediments (still continental crustal material, though now in sand-sized particles) collect in a deep trench offshore. At some time in the future, the sediments may be subducted along with the descending plate, only to be later melted and recycled again. If they escape subduction, they may be plastered directly onto the margin of the continent.

This complex scenario is played out wherever oceanic crust is subducted beneath continents. However, some modern subduction zones do not involve continents, but instead oceanic crust is consumed beneath another plate of ocean crust. In these settings, melting also occurs; however, rising basaltic magma is not halted by a less-dense lid of continental crust and so continues to the surface, creating volcanic arcs like the Aleutians. Such island arcs are made mostly of basalt, though partial crystallization of the magma en route may cause the remaining liquid to become more silica-rich. Island arcs can later be welded onto the edges of continents as they are rafted about in moving plates, and thus add to the continental crust.

The modern continental crust of the Andes is clearly more silica-rich than that formed in the Archean. Before there was much continental crust, there was nothing to stop basalts from reaching the surface, so Archean crust contained basalts in greater abundance. Once the continents had developed, however, they began to act as filters for ascending basaltic magma. Waylaid magmas forced repeated melting of the lower crust in a seemingly endless cycle of renewal. In effect, the Earth has refined its crustal veneer by producing magmas with ever-greater silica contents and more extreme proportions of incompatible elements.

A CRUSTY OLD PLANET

The great cellist Pablo Casals, when asked why he kept practicing so hard even after he had reached his mid-eighties, replied, "I have a notion that I am making some progress." If the formation of continental crust can be considered a sort of evolutionary progress, then the Earth, too, exhibits a dogged persistence. It has irreversibly transformed a tiny fraction, only a third of 1 percent, of its mass into continental crust, a process so incredibly inefficient that it has taken nearly 4 billion years to arrive at even this minuscule total.

The earliest continental crust, now comprising the surviving Archean cratons, formed by direct melting of mantle material, probably the keels which still undergird them. The composition of the Archean crust was diverse, composed of both basalt and granite. The creation of continental crust in the modern, Phanerozoic world has mostly been accomplished through the agency of plate tectonics, as igneous rocks formed earlier at the axes of spreading seafloors have been recycled into the mantle and been remelted during subduction. These basaltic magmas rise and underplate the overlying continents, mixing with the lower crust and causing it to melt. The resulting granitic magmas then ascend into the upper crust. Over eons, the continental crust has been refined into rocks with densities so light that they can never again be subsumed.

No other planet has continents composed of silica-rich rock. Other planets have crusts, to be sure, but as far as we know, their compositions are basaltic, very different from those of our world's continents. Venus, so similar in size to the Earth, has high-standing regions that look superficially like continents, but they appear to be mostly crumpled basalt. Mars and Mercury, too, seem to have veneers of basalt everywhere. The Moon has a thick crust of feldspar that floated to the top of its magma ocean, but the composition of this magma was also basaltic.

Our world is also unique in the continual rearrangement of its continental crust by plate tectonics. Rafts of solidified siliceous rock, in the form of volcanic arcs, have been periodically plastered onto the edges of older cratons, and at other times the amalgamated continents have been ripped apart, so that their fragments wandered anew over the globe. The distinctive granites of the Earth's continental crust may owe their existence to subduction and to water, which act together to produce these unusual magmas. Our planet's restless platforms of ancient siliceous scum have no counterpart, at least in this solar system.

The familiar continents—the rocky substrate on which we exist, the source of useful stones with which we construct the buildings and highways of our civilization, the host of virtually all our raw materials—are thus revealed to be a planetary oddity. They are yet another example of the marvelous workings of a unique and geologically evolving world.

Some Suggested Readings

Brown, G., Hawkesworth, C., and Wilson, C., eds. *Understanding the Earth* (1992). Cambridge University Press, Cambridge, UK. *This is a superb, though somewhat technical set of readings on various geologic subjects. Chapter 8 on "The Continents" by K. O'Nions discusses the composition, age, and origin of continental crust, and Chapter 10 on "Plate Tectonics and Continental Drift" by A. G. Smith provides a synopsis of current thinking about how plates move.*

Continents Adrift and Continents Aground: Readings from Scientific American (1976). W. H. Freeman and Company, San Francisco. *This highly readable though somewhat dated collection of readings describes the state of understanding of plate tectonics just after the revolution had fully taken hold. Especially useful are selections by A. Hallam on "Alfred Wegener and the Hypothesis of Continental Drift," J. T. Wilson on "Continental Drift," and R. S. Dietz and J. C. Holden on "The Breakup of Pangaea."*

Cox, A., and Hart, R. B. (1986). *Plate Tectonics—How It Works.* Blackwell

Scientific Publications, Oxford, UK. *A thorough treatment of modern plate tectonics.*

Dalziel, I. W. D. (1995). Earth before Pangaea. *Scientific American,* vol. 273, no. 1, pp. 58–63. *A modern view of the difficulties in reconstructing the assembly and breakup of supercontinents.*

Oreskes, N. (1988). The rejection of continental drift. *Historical Studies in Physical Sciences,* vol. 18, no. 2, pp. 311–48. *A critical review of Alfred Wegener's role in the revolution of geology.*

Taylor, S. R. and McLennan, S. M. (1996). The evolution of continental crust. *Scientific American,* vol. 274, no. 1, pp. 76–81. *This clear, beautifully illustrated article summarizes modern thinking on the growth of continental crust over geologic time.*

Van Andel, T. H. (1985). *New Views on an Old Planet: Continental Drift and the History of the Earth.* Cambridge University Press, New York. *This interesting book describes, in rich detail, the role that plate tectonics has played in Earth history.*

High-Water Mark

Origin of the Oceans and Atmosphere

RING AROUND THE SEA

Each summer my boat is docked at a landing on the Tennessee River, a quarter mile from my house. When I haul it out of the water in the fall, its bottom and sides are always encrusted with nasty gunk. Even vigorous scrubbing won't remove the last traces of the most persistent stain, the line marking the level at which the boat floats in the water. A familiar example of the same phenomenon is a bathtub ring, which delimits the water depth when the tub was last filled. On a grander scale, a flood leaves clear marks of its crest on the sides of buildings, rocks, and trees. Perhaps it is not surprising, then, that ocean transgressions in the geologic past have also left high water marks. As the sea lapped over the land's edge, it formed a new coastline defined by dumped sediments. To cite but one example, the maximum incursion of the Cretaceous ocean onto the North American continent is now revealed on the Atlantic seaboard by deposits of sand, the western margin of the Atlantic Coastal Plain.

Similar high water marks on all the continents reveal that at various times in the Earth's long history, the oceans have contained more water than they do now, usually a result of warmer climates that melted the polar caps. High water marks can also occur even if the amount of liquid water is unchanged, by shrinking the sizes of its containers or, in other words, by lessening the capacities of the ocean basins. A faster seafloor spreading rate produces a greater proportion of elevated mid-ocean ridges and thus raises sea level. But if we ignore changes in the physical state of water wrought by climate fluctuations and sea-level oscillations due to varying tectonic rates, we would find that our present world is approximately at its high water mark; that is, the Earth's total inventory of surface and near-surface water, taken as the sum of all its oceans, lakes, rivers, and groundwater, plus polar ice and atmospheric water vapor, is probably somewhere near its maximum value. Did it reach this value instantaneously, or was there a time when the oceans were mere puddles?

There is obviously an important connection, an interplay really, between the oceans and the atmosphere, as water restlessly cycles back and forth between these liquid and gaseous reservoirs. Like the ocean's high water mark, the boundary between the atmosphere and the Earth's solid surface is etched by lasting markings. Just as exposed metal parts on my boat corrode when they react with oxygen in the atmosphere, many minerals in the crust, once exposed to oxygen in the air, react to form new phases. But the atmosphere has not always had the same composition and, most important for us, its carbon dioxide content has decreased and its oxygen content has increased considerably over geologic time. To extend the analogy, I suppose we might say that, based on the extent of corrosive reactions we see in surface rocks, the present atmosphere must be near its high oxygen mark.

It is nice to think that there is no shortage of water to drink and oxygen-laden air to breathe, but it has not always been so. The early Earth was bald, probably resembling the arid and airless Moon. So how and when did the ocean-atmosphere system, this sloshing and muggy blue shroud that makes life possible; come to be? This is a difficult question to answer, because seawater and the air do not retain the same kind of indelible geologic record as does the land. There is a memory of sorts, but it is written in chemistry, not stone.

WHENCE WATER?

When you stand at the seashore and gaze out over the vast ocean, you are seeing one of the most bizarre phenomena in the solar system. No other planet, not a single one, has any liquid water splashing about on its surface, much less enough to drown three-quarters of its area. Half a century ago, we presumed that this strange substance, so plentiful on Earth, must occur everywhere. Mars was popularly thought to have canals, a maze of irrigation ditches carrying water from the poles to thirsty populations of Martians in equatorial cities; however, *Viking* spacecraft revealed the planet's surface to be dry and mostly airless, devoid of extant life and the precious fluid that sustains it. Similarly we imagined, hidden under its creamy yellow cloud banks, a humid, swampy Venus, possibly populated with jungles and dinosaurs; but measurements of its blistering surface temperature and vivid radar images of its harsh volcanic surface indicate that it too is bone dry and lifeless. Planetary exploration has profoundly changed our expectations of the habitability of other worlds, and future colonists had better be prepared to truck in their own water or find clever ways of extracting it from the interior.

So what is so special about the Earth that it is awash in water? It seems implausible to think that our planet was singled out as the only one to receive water during its formation. The answer to this riddle actually involves several issues: how the Earth acquired its water in the first place, and how it managed to retain the water over its long history. Of course, the question should really be addressed in broader terms, because we are not only interested in water, but in all of the volatile elements that are part of the oceans, rivers, and polar caps (sometimes collectively called the *hydrosphere*) and the atmosphere. These volatiles include, in addition to the hydrogen and oxygen that comprise the water, other abundant elements like carbon, nitrogen, and sulfur in various combinations with hydrogen and oxygen, as well as rarer gases such as helium, neon, argon, xenon, and krypton.

The Earth accreted piecemeal from planetesimals. Once our fledgling planet had achieved a certain size and mass, its gravity would have begun to attract gas molecules from the surrounding nebula. In fact, the planet's mass is sufficient for it to have captured a lot of atmosphere in this way, enough to generate a surface pressure a thousand times greater than the one we presently enjoy. Could our current atmosphere be a mere vestige of this primordial one, and our present oceans the condensate from this massive gaseous envelope?

Knowing that the solar nebula had hydrogen, oxygen, carbon, sulfur, and other elements in the same proportions as does the Sun, we can calculate what kinds of molecules they should have formed as the gas cloud cooled. An atmosphere snatched directly from the solar nebula would have been composed mostly of hydrogen—so much hydrogen, in fact, that it would have reacted with all the other common elements. Combined with carbon it must have formed methane, with nitrogen ammonia, with sulfur hydrogen sulfide, and with oxygen water vapor, resulting in a choking haze very different from what we now breathe. Before we scratch our heads from trying to find a way to derive our present atmosphere from this smelly mixture, let's first see whether it is even necessary.

We can test the plausibility of captured nebular gas by examining the current atmosphere's abundances of *noble gases*. These elements—neon, argon, krypton, and xenon—are called "noble" because they disdain to form compounds with other, less exalted chemicals. The fact that they are unreactive with the crust means that, once captured from the nebula, they would have remained secluded in their lofty castle, the atmosphere. Moreover, they are such heavy atoms that the planet's gravity grips them very tightly, minimizing the possibility of their escape into space. The point is this: however the early atmosphere might have changed over time, nothing much would have happened to its noble gases. Measurements of the relative proportions of noble gases in the Earth's atmosphere, as well as in the atmospheres of Mars and Venus, closely resemble each other, but not those in the Sun. Of particular importance is the fact that atmospheric neon is hundreds of times less abundant, relative to other noble gases. If the early Earth had captured nebular gas, which by its very definition had solar elemental abundances, then its neon content would be much greater than we see. Furthermore, the abundances of the noble gases other than neon in the present atmosphere are all far too low to have been snatched during reactions between a hot, early Earth and the nebula. Clearly, any primordial nebular atmosphere that the world once may have had is long gone, possibly blown away by massive collisions during accretion or stripped off during the Sun's violent T Tauri phase.

If the Earth did not get its present water and other volatiles directly by capturing nebular gas, we must find some other source, one that came into play later, after the primordial nebular atmosphere was lost. The nature of this source is suggested by the observation that the relative abundances of noble gases in planetary atmospheres are similar to those in chondritic meteorites. Could our planet's volatiles have been carried in on the backs of accreting planetesimals? At first glance this seems reasonable, but on further reflection, there is a problem. The Earth was assembled mostly from differentiated planetesimals, small bodies that were themselves already depleted in volatiles. A way around this problem might be to accrete, almost at the last minute, some chondritic planetesimals, forming a thin, volatile-rich shell on the almost-finished Earth. Let's see if that might do the trick.

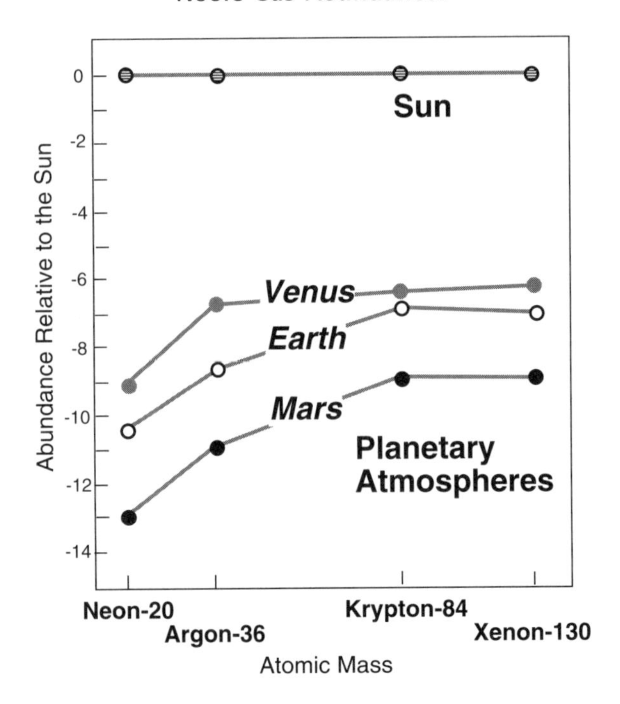

The abundances of noble gases in the atmospheres of Earth, Mars, and Venus are very different from solar abundances. This figure illustrates the measured concentrations of nonradiogenic isotopes of planetary neon, argon, krypton, and xenon, divided by those in the Sun, shown at the top. Because the Sun contains many more atoms of gas than a planetary atmosphere, all of the abundances are relative to one million atoms of silicon. The differences between the various gas reservoirs shown here are so great that the vertical scale is logarithmic.

Some kinds of chondrites, notably carbonaceous chondrites, contain as much as 20 percent water, so they could be a reasonable source for the world's oceans. Other observations shore up the idea that such volatile-rich chondritic matter might have been added to the Earth late in its accretional history. Judging from asteroid spectra, the carbonaceous chondrites originated far past the orbit of Mars, where nebular temperatures were cold enough for ice to have condensed. In some of these bodies the ice melted and formed hydrous minerals, while in others it was preserved intact. This was a very different situation from the inner belt, where asteroids were differentiated and contained little or no water. However, after the dry bodies in the inner belt differentiated, their surfaces were pelted with carbonaceous chondrites. We know this because small fragments of these volatile-rich chondrites occur commonly as

This grapefruit-sized igneous meteorite (an achondrite), which fell in Cumberland Falls, Kentucky, in 1919, contains black clasts of carbonaceous chondrite rich in volatile elements. The surface of the parent asteroid of this meteorite was once pelted with such material. Similar accretion of a late veneer of carbonaceous chondrite onto the early Earth may have brought in water for the planet's oceans. (Smithsonian Institution)

clasts in regolith (soil) samples from differentiated asteroids, but not in samples from their interiors. What happened to these asteroids might logically have also happened to planets, but on a grander scale. We might speculate that waves of carbonaceous chondrites slowly filtered into the inner asteroid belt, and thence into the inner planet region. Working their way inward from the the far-flung, frozen reaches of the nebula, these dark boulders, bearing precious water, only reached the Earth's feeding zone during the waning stages of accretion. Those that slammed into the newly formed planet painted its surface with a veneer of volatile-rich matter.

Yet another potential source, perhaps even a better source, of the Earth's water was comets, which are thought to reside in huge quantity in the outer reaches of the solar system. Comets are dirty snowballs, consisting mostly of water and other volatiles with a sprinking of mineral matter. If only a tenth of the objects bombarding the Earth between 4.4

and 3.8 billion years ago were comets, they could have provided all the water needed for the present oceans. These icy lumps would also have brought in considerable amounts of carbon dioxide and other volatile species.

It is not exactly clear what happened next. Impacts are violent events, and some volatiles would certainly have been lost during infall. The gases that the Earth managed to retain would have formed a steam atmosphere under the infernolike conditions of planetary accretion. As the hailstorm of rocks and ice subsided, water vapor might have condensed directly to form oceans, but significant amounts must also have dissolved in the underlying magma ocean. When this roiling sea of molten magma finally cooled and solidified, a lot of water and other volatiles were sequestered inside the solid planet (in rocks that now comprise the crust and outer mantle), possibly to be released at some later time. Carbon dioxide and nitrogen are much less soluble in magma than is water, so these gases were probably concentrated in the early atmosphere.

GRADUAL OR INSTANT OCEANS

As anyone who has witnessed a volcanic eruption can testify, volcanoes spew forth huge quantities of noxious gases. In 1951, W. W. Rubey of the U.S. Geological Survey suggested a connection between volcanoes and surface volatiles. He argued that the volatiles that comprise the atmosphere and hydrosphere had been released from the mantle during volcanism, a process that came to be known as *outgassing*. Rubey envisioned outgassing occurring gradually over a time span measured in billions of years, and probably still going on at the present day. His model implied that sea level should have risen slowly and progressively over geologic time. It is also conceivable, though, that the oceans formed almost instantaneously during rapid outgassing of the early Earth, an argument put forward in 1971 by Fraser Fanale, then at the California Institute of Technology. How can we choose between these alternatives?

Helium (yet another noble gas) is such a light atom that it is used in filling blimps and party balloons; in fact, it is so light that the Earth's gravity cannot hold it for long, and so atmospheric helium is lost to space. As a consequence, the helium gas in the present atmosphere, unlike other heavier noble gases, has not been there very long. Its atmospheric concentration, then, depends on the rate of its escape to space versus the rate at which it is replenished by outgassing of the Earth's interior.

Helium is actually a mixture of two isotopes. Helium-4, the more abundant one, is continually created as a by-product of the radioactive decay of uranium and thorium isotopes. The amount of radiogenic helium-4 escaping from the atmosphere seems to be approximately balanced by the amount produced by radioactive decay in the crust. If only helium-4 were leaking out of the Earth, there would be no need to postulate mantle outgassing at all, because this isotope did not exist long ago when volatiles were originally trapped in the magma ocean. The other isotope, helium-3, is not a product of radioactive decay, so for many years it was unclear why it was present in the atmosphere. This mystery was finally resolved in 1969 when scientists discovered small quantities of helium gas, laden with a higher than normal proportion of helium-3, in seawater near mid-ocean ridges. This helium is apparently being continuously outgassed from the Earth's mantle during the eruption of basaltic magma onto the seafloor. The exhaled helium-3 atoms really are gaseous fossils, matter that was once added to the Earth by the accretion of volatile-rich planetesimals or comets and that is only now being released.

By measuring the tiny whiffs of helium-3 in volcanic regions all over the world, it is possible to estimate the rate at which this element is outgassed. Furthermore, the degassing rates for other noble gases can also be determined by comparing their amounts in volcanic gases relative to helium-3. In this way, we can see just how rapidly, or slowly, the Earth is purging its interior of the unwanted noble gases. The bottom line is that the present rate of planetary outgassing is far too slow to account for the large amounts of noble gases in the present atmosphere. Instead, these data demand that a significant part of the atmospheric noble gases, perhaps as much as 80 percent, must have been outgassed rapidly in some long-vanished era of the planet's history. Presumably, water for the oceans was exhaled at the same time.

So, both Rubey and Fanale were partly right and partly wrong. Outgassing of the interior has occurred over most of our planet's lifetime, but the bulk of the volatiles in the hydrosphere and atmosphere must have been released suddenly in some early event, perhaps liberated from accreting planetesimals during infall or burped out of the magma ocean.

ALL THAT SALT

As any swimmer who has inadvertently swallowed a mouthfull of seawater knows all too well, the ocean tastes awful. The reason is that about 3 percent of seawater is dissolved minerals, the bulk of which is common table salt, sodium chloride. But its bitter flavor is also heightened by quantities of dissolved magnesium, calcium, potassium, and a host of other elements. The chemistry of the ocean may not do much for its taste, but it can tell us a great deal about its history.

More than three hundred years ago the British scientist Robert Boyle first analyzed small concentrations of salt in rivers flowing to the sea. His discovery prompted the hypothesis that the salt in the oceans was derived from the land. Rain falling on the continents slowly dissolved rocks and, when gathered into rivers, carried this material ultimately into the ocean. As seawater evaporated to form clouds, it left behind its dissolved constituents and so became progressively saltier with time. This picture of rivers as conveyor belts for salt and other dissolved substances was the accepted scientific view for centuries, and estimates for the rate of salt addition to the oceans formed the basis for calculations of the age of the seas and of the Earth.

There are serious problems with this idea, however. For one thing, the amount of salt carried in rivers is sufficiently large that the oceans would have reached their present salinity in a surprisingly short time, less than 100 million years. For another, there is no evidence that the seas have become saltier with time. Direct descendants of early life forms still inhabit the oceans, from which many biologists have inferred that there has been no drastic change in the marine chemical environment. Moreover, body fluids for most organisms contain concentrations of salts similar to that in seawater, and there is no reason for them to have uniform salinity if they evolved at different times in oceans that were becoming ever more salty.

So, the sea cannot be viewed simply as a large, inert kettle into which dissolved constituents are dumped, to remain forever. Instead, it must be a vast chemical recycling bin, accepting salt and other materials while, at the same time, somehow removing them in insoluble forms. Not all elements are affected by the same processes. Some, like calcium and silicon, are incorporated into the skeletons of living organisms, later to sink to the ocean bottom when the creatures no longer need them; whereas others, like potassium, react to form clay minerals on the ocean floor. The rate of biologic or geologic processing is different for each element, leading to varying times of residence for each dissolved constituent in seawater. Aluminum, for example, has a very short average residence time of only a few decades, whereas calcium persists for about a million years and sodium remains in the oceans, on average, for 68 million years. The residence time of chlorine is very long—for the purpose of this discussion, forever. Thus the sea tastes salty because sodium chloride has a longer residence time than most other dissolved constituents.

The idea that rivers feed the sea all its dissolved constituents has been a useful one, but recent discoveries have now drastically altered this perspective. There is a serious mismatch between the heat predicted to be released from mid-ocean ridges and the flow of heat actually observed there, amounting to a shortfall of about 40 percent. The calculations seem to be sound, because theory and observations agree on cooler oceanic crust away from the ridge crest. A huge amount of the Earth's liberated heat is unaccounted for in these models, heat that must be eliminated by some other means than conduction through rocks. One way of doing this is by circulating cold water through rocks at mid-ocean ridges; a technology based on the same principle is commonly used to cool automobile engines. Pulling apart an oceanic spreading center causes it to be riddled with cracks, conduits that allow cold seawater to circulate down into the hot crust. This water, once heated, migrates back up to the surface and escapes via submarine hot springs into the overlying ocean. Active hot springs have now been discovered on the ocean floors near mid-ocean ridges.

The importance of this circulation of seawater lies in the opportunity it affords for chemical interaction. Some elements are dissolved from the rocks and added to seawater, while at the same time others are removed from the water by reactions with hot rocks. For example, sodium and magnesium in seawater are lost, and so their concentrations are prevented from increasing to extreme levels; in contrast, the iron and manganese abundances in seawater are increased. The change in the chemistry of the water is manifested by what happens as it emerges from submarine hot springs. It is often so laden with dissolved metals that they immediately precipitate on contact with the cold ocean. These vents build up pipes of precipitated sulfides, sometimes resembling smokestacks that belch black soot. A huge amount of water is circulated through the crust in this way—the equivalent of the entire ocean volume flows through mid-ocean ridges every 10 million years or so. Thus, the chemical composition of seawater may be largely controlled by its in-

Submarine hot springs, called "black smokers," are found at a depth of several kilometers below sea level on the East Pacific Rise. The water erupting from the vent has a temperature of over 300 degrees Celsius. It is actually clear, but fine, black particles of iron sulfide and other minerals precipitate when the hot water comes in contact with the cold ocean. The chimneylike structure is built up from precipitating sulfides.

teraction with rocks of the ocean floor, rather than regulated by whatever rivers discharge into it.

A GREENHOUSE UNDER A PALTRY SUN

We know that the Archean Earth had liquid oceans, because the common occurrence of sedimentary rocks from that era required water for the weathering, erosion, transport, and deposition of sediments. Liquid water exists only in a narrow temperature range (0 to 100 degrees Celsius), broadly defining climates that we might describe as equable. Not every part of the planet had to experience these climatic conditions, of course, but temperatures over most of the Archean world must have been bracketed between the freezing point and boiling point of water.

That is actually something of a surprise. During Archean time, the Sun was significantly less luminous than now, a fact well appreciated by astrophysicists but virtually ignored by geologists until just a few decades ago. As hydrogen is continually fused into helium in the core of the Sun, the mean atomic weight of the atoms increases (helium is four times heavier than hydrogen). This causes a commensurate increase in the density of particles in the Sun, which promotes faster collision of hydrogen atoms and thus quickens the rate at which they fuse together. Faster fusion equates to a hotter Sun in modern times, but during the Archean the solar luminosity was perhaps only three-quarters of its present value. The lower amount of sunlight impinging on the Earth should have resulted in temperatures so cold that all the oceans froze solid. But that obviously did not happen.

To understand what compensated for the diminished solar heat during this early part of the Earth's history, we must see how temperature is related to atmospheric composition. Carbon dioxide must have been very abundant in the Archean atmosphere, because it was much less soluble in the magma ocean than was water. This molecule has a curious property, perhaps best illustrated with an analogy. If you park your car in the sun on a hot, summer day, with the windows rolled up, you are apt to feel a blast of hot air when you later open the door. The glass windshield is transparent to sunlight at visible wavelengths, but not at infrared wavelengths. Once the visible light energy enters the car and hits the seats or dashboard, it is transformed into infrared photons. Because the windows act as barriers to the escape of infrared radiation, the interior swelters. A similar phenomenon occurs in greenhouses, giving rise to the description of this process as the *greenhouse effect*.

For a planet, the role of the greenhouse window can be played by carbon dioxide in the atmosphere. Visible light reaching the ground is transformed into infrared radiation, which is then captured by vibrating carbon dioxide molecules. (Water is also an excellent greenhouse gas, but the amount of water in the Archean atmosphere was much less than the amount of carbon dioxide.) At the present time, traces of water (less than 1 percent) and carbon dioxide (0.03 percent) in the Earth's atmosphere are responsible for increasing its surface temperature by a small but comfortable amount, though there is legitimate concern that the carbon dioxide our industrialized society pumps into the air may promote too much warming. Because the atmosphere of the early Earth contained a huge amount of carbon dioxide, heating was correspondingly greater, apparently sufficient to offset the effect of a fainter young Sun. To obtain temperatures in the necessary range requires that the atmosphere was about half carbon dioxide, with the remainder nitrogen and perhaps a pinch of water vapor.

JUST RIGHT

Where did all this carbon dioxide go? The present atmosphere is mostly nitrogen and oxygen, with only a trace of carbon dioxide. Ocean water contains a little dissolved carbon dioxide, but it amounts to only a tiny fraction of what must be missing. Returning to the idea of the ocean as a chemical factory, though, we can explain the vanished carbon dioxide. When silicate rocks on the continents are weathered, they release calcium and magnesium which are transported to the ocean. There these atoms react with the dissolved carbon dioxide to form calcium and magnesium carbonates, which then precipitate onto the ocean floor to form limestones. (In the modern world, the limestones are produced mostly by shell-forming organisms, but the process is chemically similar to inorganic precipitation. In the Archean era, limestones could have precipitated directly or formed by the action of primitive bacteria.) Once carbonate is removed, the ocean extracts more carbon dioxide gas from the atmosphere to replace what it just lost and, as the process continues, the atmosphere is progressively depleted in carbon dioxide as it is locked up into rocks on the seafloor. This process is remarkably, even embarrassingly, quick. Calculations indicate that virtually all of the carbon dioxide in the atmosphere could have been removed in less than a million years. So why is there any carbon dioxide left at all? To make up for the lower output of the Archean Sun and prevent the oceans from freezing, carbon dioxide had to have persisted as an important constituent of the atmosphere for billions of years. Somehow carbon dioxide in rocks must have been returned to the atmosphere.

Plate tectonics accomplishes that feat today. Limestones on the seafloor may eventually be subducted into the mantle, where they are heated. Metamorphism causes the carbonates to recrystallize into new minerals and, in the process, they lose carbon dioxide. This gas eventually makes its way again to the surface and is released back into the atmosphere. Even though limestones that are parts of land masses are not subject to subduction, they also can release carbon dioxide. During the collision of continents, such as the one between Asia and India that formed the Himalayas, the metamorphism of carbonate rocks inside uplifted mountain belts releases so much carbon dioxide that it may spur episodes of global warming.

In the early Archean, plate tectonics may not yet have been fully under way, but some kind of recycling of crustal rocks clearly took place. By whatever means, carbonate rocks must have been destroyed and lost

The Carbon Cycle

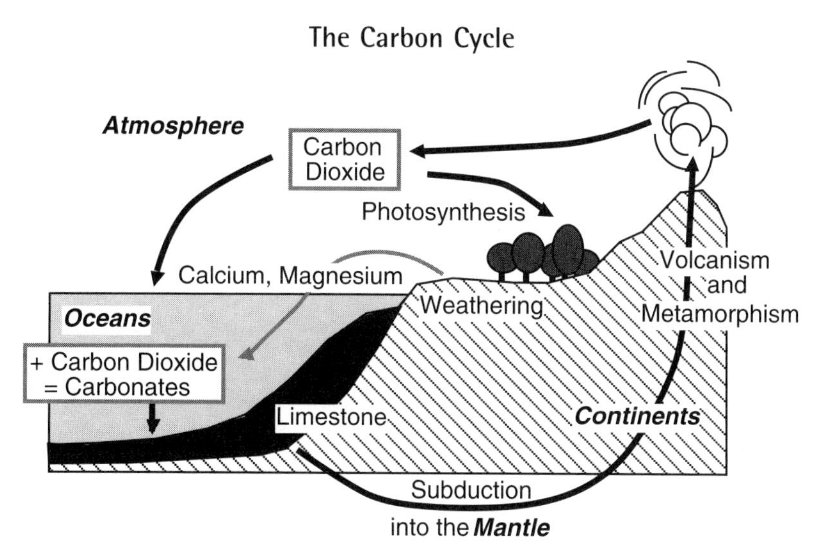

The carbon cycle describes the continuous movement of carbon through the Earth's atmosphere, hydrosphere, and solid interior. Carbon dioxide is released into the atmosphere during volcanic eruptions. Some of this material is then dissolved into the oceans, where it reacts with calcium and magnesium liberated from the continents during weathering. The carbonates thus produced form limestone on the ocean floor. At subduction zones, limestones are metamorphosed, releasing carbon dioxide which is reintroduced to the surface during volcanism. Green plants also regulate carbon dioxide abundance in the atmosphere by removing it during photosynthesis.

their carbon dioxide, allowing the grand carbon cycle, from atmosphere to oceans to rock, and back again, to occur.

In a famous story, Goldilocks settled on a bowl of porridge that was just right, not too hot and not too cold. The not-too-hot, not-too-cold Earth clearly required some kind of thermostat to regulate its temperature excursions over billions of years, given the continuously increasing solar luminosity. The carbon cycle has an important feedback mechanism that apparently serves as a brake on temperature changes. When the Earth is warmer, due to high atmospheric carbon dioxide, rock weathering reactions are faster because of increased rainfall. Thus, carbon dioxide is incorporated into limestones on the ocean floor more rapidly. Because the rate of return of carbon dioxide to the atmosphere by tectonics is not affected by climate, the net effect is the removal of atmospheric carbon dioxide and the lowering of temperature. Conversely, as the climate becomes colder, weathering slows and the amount of limestone produced in the oceans decreases. Increasing carbon dioxide in the atmosphere then causes warming.

The Goldilocks phenomenon seems to have occurred throughout most of Earth's history, but there have been a few intervals where temperatures apparently oscillated to extremes. During the Proterozoic era, for example, there were a number of long and rather closely spaced ice ages. During this time, deposits of glacial sediments formed even at middle and low latitudes, so it is doubtful that these cold periods can be simply attributed to their host continents migrating across the poles. Rather, it seems likely that the world was locked in a global freeze. With few exceptions, however, the Earth's temperatures have been held to modest values throughout its history.

This miraculous planetary thermostat is absent on our planetary neighbors. Venus has no oceans of water, so it has no way of tying up its atmospheric carbon dioxide in rocks. Its massive gas envelope (ninety atmospheres of mostly carbon dioxide) is a greenhouse run amok, with surface temperatures approaching 500 degrees Celsius. Mars also has no oceans and only a thin atmosphere. Abandoned river channels in the ancient Martian highlands suggest that early in its history it had surface water, but lack of plate tectonics prevented any limestones that might have formed from later yielding up their greenhouse gases.

WATER'S ROUND TRIP

Because the noble gases outgassed from the Earth's interior have steadily accumulated in the atmosphere, we can consider the path from mantle to atmosphere for these gases to have been a one-way trip. That is not the case, however, for most volatiles. We have already seen how carbon dioxide is recycled, from atmosphere to oceans, thence into carbonate rocks, and finally back into the atmosphere again.

The path of water is more difficult to trace. Most of the outgassed water has been concentrated in the oceans, and only a tiny fraction exists as atmospheric water vapor. The water molecule itself is too heavy to escape the planet's gravitational pull, but water vapor in the air is sometimes broken into its constituent atoms by ultraviolet rays. The resulting hydrogen is readily lost to space, which is why there is so little hydrogen in the present atmosphere. Hydrogen is just too light for the Earth's gravity to hold on to it. The net effect is the gradual loss of water vapor in the air, which is then replaced by evaporation from the oceans. In this way the Earth loses an amount of water equivalent to the volume of a small lake each year. Over the planet's history, however, this amounts to less than 0.2 percent of the water in the oceans, only a minuscule fraction of its total inventory.

At one time it was accepted that outgassed water, like noble gases, had only a one-way ticket to the surface. However, there is no reason why water could not be cycled like carbon dioxide. Reactions of seawater with rocks of the oceanic crust cause water to be incorporated into hydrous minerals. When these are subducted deeply enough and metamorphosed, the water is released. Water has the curious property of causing mantle rocks to melt at lower temperatures than they would otherwise. The magmas generated in this way dissolve the water and carry it again to the surface, where it is emitted into the atmosphere. It is estimated that perhaps a fourth of the world's water resides in rocks of the crust, and much of this may be recycled.

AN IMPROBABLE ATMOSPHERE

Gaseous oxygen, comprising 21 percent of the Earth's atmosphere, is an odd commodity. This element, in its free state, is present nowhere else in the solar system in such abundance. By all rights, it should not be here either. Left to its own devices, oxygen in the air would react with surface rocks; it would be fully consumed within about 6 million years, and the world would revert to an anoxic state. Atmospheric oxygen leads a precarious existence, one made possible only by the intervention of green plants. Life has made possible this bizarre shroud of breathable gas, and in consequence has itself been drastically transformed.

The Hadean and Archean Earth had a very different atmosphere, composed mostly of carbon dioxide and nitrogen, as we have already seen. It also contained hydrogen sulfide, a little water, and perhaps some methane and ammonia, a noxious mixture that most current life forms would view with alarm. The absence, or at least very low abundance, of oxygen from this mix is revealed by the peculiar nature of ancient ore deposits.

The Witwatersrand basin of South Africa is an important source of gold and uranium. The ore is part of a conglomerate of quartz pebbles, a rock unit that is very old, perhaps 2.3 billion years. Within this sediment are grains of gold and uraninite (uranium oxide), minerals that were broken out of older rocks as they weathered, carried downstream by rivers, and finally dumped as placer deposits. Explaining the gold is no problem; it is an inert mineral that readily survives weathering and

The formation of this banded iron formation from Minnesota suggests that the Earth's atmosphere contained much less oxygen several billion years ago than it does today.

transport in rivers. What makes this ore so curious is the fact that uraninite is highly vulnerable to oxygen and is destroyed under present atmospheric conditions. It is difficult to see how the uraninite could have been weathered out of rocks and transported to the site of deposition without having intimate contact with the Archean atmosphere. The fact that uraninite occurs in ancient sediments tells us that, before 2 billion years ago, the air contained much less free oxygen than exists now.

The world's largest concentrations of iron ore occur in the Hamersley Range of Australia, the Krivoy Rog of Russia, the Labrador Trough of Canada, and Minas Gerais in Brazil. All of these valuable deposits are banded iron formations, composed of alternating layers of iron oxide and chert. They are gigantic, and they are old, all formed between 2.0 and 2.5 billion years ago. The minerals of banded iron formations are thought to be precipitates that rained out onto the floors of Archean seas, but this mode of formation poses a problem: To precipitate from seawater, iron must first be dissolved. That does not happen in today's world. In the presence of free oxygen, iron oxidizes to form ferric iron, which is insoluble. If the Archean atmosphere contained less free oxygen, however, iron would have occurred in the ferrous state, a condition in which it could have dissolved in seawater.

Taken together, the occurrence of uraninite-bearing sediments and banded iron formations, known only from Archean time, indicates that oxygen may be a relatively recent addition to our atmosphere. Prior to about 2 billion years ago, the oxygen content of air was probably only about 2 percent of its present value. Iron-bearing sediments formed after that time reveal an increase in free oxygen. Called "redbeds" for their distinctive coloration caused by hematite (highly oxidized, or rusty iron), these rocks formed when iron was weathered out of rocks in the presence of enough oxygen to convert it to the ferric state.

Oxygen is a terribly reactive substance, as illustrated by its central role in combustion. Its chemical combination with iron and other elements proceeds relatively rapidly even in the absence of conflagration. The presence of free oxygen as an atmospheric substance requires that its potential reactants first had to be satisfied. Only after virtually all the iron on the surface had finally been laid to rest as ferric oxide, all the carbon transformed to carbon dioxide, all hydrogen to water, all sulfur to sulfate, and any methane or ammonia to more oxidized compounds, could oxygen begin to accumulate. All this took a considerable amount of oxygen, so something had to be pumping this element into the atmosphere for a long time.

The earliest source of oxygen was the breakdown of water vapor by ultraviolet sunlight, a process known as *photodissociation*. We have already mentioned this process as a loss mechanism for water, because the liberated hydrogen atoms escape to interplanetary space. The oxygen atoms are heavier, though, and so remained in the Archean atmosphere long enough to react with other elements. At the present time, this process generates several million tons of raw oxygen each year. The capacity of the oxygen sinks discussed above is so great, however, that photodissociation, acting alone, would require another 26 billion years to allow oxygen to rise to its current atmospheric level. The process is also self-limiting; some of the free oxygen combines to form ozone, which is an efficient shield against ultraviolet light. Four billion years ago or more, the earliest additions of oxygen to the atmosphere probably resulted from photodissociation, but some other source was clearly needed to achieve the very high levels we have now.

That source was *photosynthesis*. Using energy from sunlight, green plants convert carbon dioxide and water into useful organic compounds, giving off oxygen as a by-product. Today this process produces oxygen in copious quantities, nearly 20 billion tons per year. In Archean time, photosynthesis would have been much less efficient. The earliest photosynthesizers were probably similar to primitive anaerobes that exist today, and these organisms are poisoned by too much oxygen. In fact, oxygen levels in the Archean may have even been limited by this intolerance. Eventually, new organisms that could tolerate higher levels of

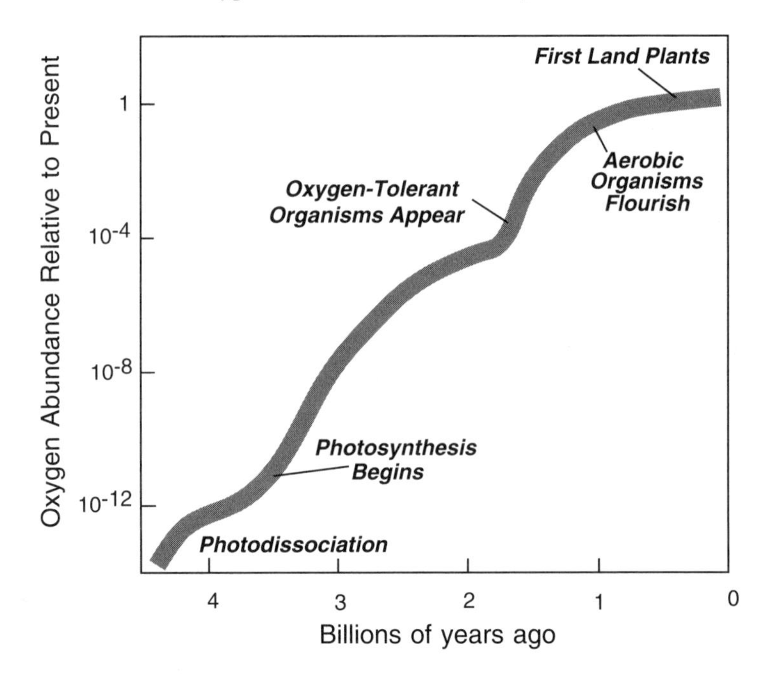

The abundance of free oxygen in the Earth's atmosphere has progressively increased over time, as organisms capable of producing and tolerating greater amounts of oxygen evolved.

oxygen evolved, plausibly about 2 billion years ago, allowing the photosynthetic production of oxygen to escalate rapidly. Once the atmosphere and oceans contained free oxygen, banded iron formations could no longer form and redbeds appeared.

A later biologic innovation was the evolution of aerobic organisms, which flourish in free oxygen. At this point, anaerobic organisms retreated to the few remaining habitats uncontaminated by oxygen, and the number and diversity of other photosynthesizers increased, along with their breathable waste product. All this biological development happened exclusively in the oceans. Finally, green plants invaded the land during the Devonian period, beginning about 40 million years ago. Photosynthesis on land presently exceeds that in the oceans by as much as a factor of two. It seems likely, then, that the global rate of atmospheric oxygen production must have increased markedly at that time.

A DAMP GODDESS

James Lovelock, a British atmospheric chemist, made a proposal in the 1960s that has since caused apoplexy in many scientific circles. Lovelock appropriated the term *Gaia,* the name of a Greek goddess, for his controversial idea. We should not read too much into his choice of name, though, as he felt this was simply "a good four-letter word referring to the Earth." The Gaia hypothesis can be succinctly summarized as follows:

> The Gaia hypothesis states that the chemical composition of the reactive gases and the temperature of the Earth's atmosphere are biologically controlled. Certain features, e.g. the salinity and alkalinity of the hydrosphere, are moderated by the biota in that their range of variation is kept within tolerable limits. . . . Gaian environmental regulation is achieved largely by the origin, exponential growth, and extinction of organisms, all related by ancestry and physically connected by proximity to the fluid phases (water and air) at the Earth's surface.
>
> —Lynn Margulis and Oona West, *GSA Today*

This imaginative hypothesis repels many scientists, but in fairness it should be noted that it has often been misunderstood. A great deal of nonscientific, one might even say bizarre, Gaia literature has been published, and press attention has focused especially on the purported Gaian view that the Earth functions as (or perhaps is) a single living being. This notion has been specifically rejected by some Gaia proponents who are scientists, based on the fact that no single organism feeds on its own waste or, by itself, recycles its own food. So the controversy (at least the informed controversy) hinges, then, on whether life regulates the planetary environment to its liking or, conversely, whether life merely responds to environmental changes. Although I only sit in the bleachers watching this debate unfold, I confess that my sympathies lie with the skeptics.

It is clear that life has exerted a profound influence in the creation of an atmosphere wildly out of equilibrium by radically increasing its oxygen content. This, in turn, has set the very course of biologic evolution on a new path. As pointedly stated by Gaian advocates, "The biosphere hums with the thrill and danger of free oxygen." Plant life also diminishes the atmospheric carbon dioxide level through photosynthesis, which moderates the planet's temperature. But is this really control, or is it response? Would the carbon cycle, which truly regulates climate, suddenly collapse if there were no life to fix carbon into limestone, or would the seas revert to inorganic precipitation to accomplish the same thing? To be sure, there is an interconnectedness between life and the environment, but the Earth has special, nonbiologic attributes that, by themselves, would have led it along a different path from those followed by the desiccated, hopeless planets that flank it.

Regardless of whether or not one accepts the Gaia hypothesis, it is irresistible to regard the Earth as a singular world shaped by a collusion of unique processes. Its volatiles, an inheritance of chondritic planetesimals or comets accreted as a veneer on the newly formed planet, have been exhaled from the interior in fits and starts to their present high water mark. It is a world now drenched in fluid, with oceans of sloshing, salty water inundating most of its surface. For eons its temperature has been just right to maintain the water in liquid form, adjusted by a thermostat of cycling carbon dioxide. Its improbable, combustive atmosphere was created and is maintained by the chemical magic of photosynthetic organisms. Familiarity with the sea and the air makes them seem almost ordinary, but in reality this soggy mask that covers the Earth is one of the most wondrous and unlikely aspects of creation.

Some Suggested Readings

Allegre, C. J., and Schneider, S. H. (1994). The evolution of the Earth. *Scientific American,* vol. 271, no. 4, pp. 66–75. *This article nicely summarizes current thinking on planetary outgassing, changes in atmospheric composition, and climate modifications.*

Cloud, P. S. (1988). *Oasis in Space: Earth History from the Beginning.* Norton, New York. *A beautifully written book describing in lucid terms, among other things, how the oceans and atmosphere came to be.*

Holland, H. D. (1984). *The Chemical Evolution of the Atmosphere and Oceans.* Princeton University Press, Princeton, NJ. *The current bible on scientific interpretations of the origin and evolution of the hydrosphere-atmosphere system; this will be difficult reading for nonscientists.*

Kasting, J. F. (1993). Earth's early atmosphere. *Science,* vol. 259, pp. 920–26. *This technical article presents a widely accepted model for the atmospheric composition of the early Earth.*

Margulis, L., and Sagan, D. (1986). *Microcosmos: Four Billion Years of Evolution from Our Microbial Ancestors.* Summit Books, New York. *A fasci-*

nating introduction to the Gaia hypothesis, by way of exploring the parallel evolution of life and the environment.

Van Andel, T. H. (1985). *New Views on an Old Planet.* Cambridge University Press, New York. *Chapters 10, 11, and 12 of this interesting and very readable book provide a thoughtful discussion of the origin and evolution of the oceans and atmosphere.*

Let There Be Slime

The First Life

TELLING THE QUICK FROM THE DEAD

Of all its wonders, the most remarkable feature of the Earth is surely its life. But what does it mean to be *alive*? All of us probably have a pretty firm idea of the distinction between what is living and what is not, but the scientific concept of life remains one of its most fundamental questions, and the answer is fuzzier than you might think. Aristotle taught that insects and worms arose from dewdrops and slime, that mice and fish were born of mud and sand, notions that seem silly now but passed for the science of his day. A philosophical approach to distinguishing between the quick and the dead regards organisms as imbued with a "vital force," a kind of divine spark, something outside the realm of science. This is certainly an appealing conceptual view to anyone who has suffered the death of a loved one. In contrast, the commonly held scientific (materialistic) view is that being alive is a natural consequence of molecular interactions and requires no role for vitalism. In this concept, life functions emerge from the complex integration of many separate chemical components. We might envision this as analogous to the automobile; its ability to corner and to move at high speed is not inherent in any of its separate parts alone.

Perhaps we can get a better handle on the question of what life is by defining the essential characteristics of living matter. No completely satisfactory list of characteristics exists, but the following basic properties seem critical: life must have the ability to generate new substances from simpler substances in its environment, and must have the capacity to replicate itself. That seems relatively straightforward, but using these characteristics it is still difficult to classify some kinds of entities unequivocally.

At what level do we need to search for the threshold of life? Many years ago, British biologist J. B. S. Haldane observed that there are about as many cells in a human as there are atoms in a cell. Since cells are alive and atoms are not, he suggested that "the line between living and dead matter is therefore somewhere between a cell and an atom." Within Haldane's defined boundaries lie the bacteriophages ("phages" for short), entities smaller than cells but still chemically complex. When isolated, phages do not appear to be alive, but they can reproduce themselves by hijacking the replicating machinery of the bacteria they infect. If they reproduce, does this mean phages are alive? They survive, as Haldane noted, "heating and other insults which kill the majority of organisms." If they do not die, does this mean that phages are not alive? Similarly, viruses are mere strands of genetic information encased in protein, matter scavenged from other living forms. Viruses cannot reproduce outside an infected, living cell. They, too, pose a problem for those who wish to assign the living and the nonliving to separate bins.

Materialism forms the basis for science's best definition of life. However, it remains a fact that nothing that a qualified observer would pronounce alive has yet been produced from scratch in the laboratory. As expressed so well decades ago by a noted anthropologist:

> The ingredients are known; they can be had on any drug-store shelf. You can take them yourself and pour them and wait hopefully for the resulting slime to crawl. It will not. The beautiful pulse of streaming protoplasm, that unknown organization of an unstable chemistry which makes up the life process, will not begin. Carbon, nitrogen, hydrogen, and oxygen you have mixed, and the same dead chemicals they remain.
>
> —Loren Eiseley, *The Immense Journey*

Well, maybe. It is certainly fair to say that the requirements for life remain an open question, as neither biochemistry nor some vital force has been convincingly demonstrated to cause the spontaneous transformation of the dead into the living. But it seems possible that the answer may not be long in coming. In 1993 at the Scripps Research Institute in La Jolla, California, Gerald Joyce and his colleagues created a molecule that was astonishingly lifelike. It was a mere snippet of RNA, one of the master molecules of the cell, synthesized in a test tube. Surprisingly, within an hour of its creation, it began to gather in the matter around it and make copies of itself. And the copies made more copies. Then, these bizarre molecules began to evolve, so that they could perform unexpected chemical tricks. These are functions that we commonly attribute to the living. Joyce admits that his replicating molecule was not alive, because it could not reproduce without a steady supply of externally manufactured chemicals. This strange experiment does, however, illustrate that artificially manufactured molecules can be startlingly lifelike.

Yet another way to address this question is to explore how life first *originated,* in the hope that understanding the historical transition from inert to alive may reveal what was, or is, necessary. If life can be traced backward through its simpler stages to some formative organism, perhaps the mysterious borderline to the animate can be understood. To some people, studying the origin of life may seem like science fiction, and to others it probably seems presumptuous to address a subject they would more properly assign to religion. Admittedly, it is difficult to place this topic wholly within the confining walls of science. However, there are legitimate scientific avenues for studying life's frontier, based on clues from the Earth's fossil record, from outer space, from laboratory experiments, and, most important, from the biochemistry of living organisms themselves. Let us see what they tell us.

CREATION IN THE ABYSS

In the year 1870, champions of the materialistic view of life's creation rushed forward to do battle with those who advocated vitalism. Prominent among these was the great British naturalist Thomas Huxley, who argued that the ocean depths would prove to be biologic refuges where time itself had been arrested. Within these zones of perpetual midnight lurked prehistoric animals, and on the seabed itself rested a carpet of primal ooze, called *Urschleim* by German scientists. This gelatinous scum had already been dredged up during the laying of the first Atlantic cable. Huxley believed that Urschleim, a formless, protoplasmic half-living mat, covered the ocean floor as a continuous sheet. He thought it to be matter marking the very transition from nonliving to alive, the primal agent from which all higher life forms had evolved. Swiss naturalist Louis Agassiz shared this view, observing that the oceanic abyss most closely matched the conditions under which life had first emerged. It was, he said, the depths of the ocean alone that could place animals under pressures like that afforded by the dense atmosphere of a young world. With this perspective, it was not hard to believe that the secret of life might be found in Urschleim, and even that life's creation might still be in process on the ocean bottom.

Alas, this intriguing idea came crashing down for the most irrefutable of reasons: Urschleim did not really exist. In 1872 the British Admiralty launched the *Challenger* expedition, an ambitious project to study the ocean depths. Huxley confidently predicted that the expedition would recover "zoological antiquities which in the tranquil and little changed depths of the ocean have escaped the causes of destruction at work in the shallows and represent the predominant population of a past age." Four years later the ship steamed home with enough new scientific discoveries to fill fifty huge volumes—but without Urschleim. One of the expedition's scientists had found that he could produce all of the properties of this bizarre slime by simply adding alcohol to seawater. (The original specimen Huxley studied had been preserved in this way before he examined it.) Precipitation of calcium sulfate in the treated mixture resulted in a kind of ooze that resembled protoplasm. Huxley took the news with good grace, but it was clearly a defeat for materialism.

It was not a fatal blow, however. Modern science has resurrected the concept of materialism, based primarily on research dealing with the origin of life. This was inevitable, I suppose, because the alternative, vitalism, lies outside the realm of scientific inquiry. Nevertheless, the materialistic view, though certainly not yet proven, has a number of observations to recommend it. Those who pursue the search for life's margin have found no clear demarcation between the living and the nonliving. Instead, there is almost a blurred continuum between carbon molecules of purely chemical origin and living cells. Once upon a time Urschleim, or something much like it, probably existed. And, in a marvelous twist of fate that so often characterizes scientific inquiry, it now appears that this fledgling life form may have been born in the abyss.

WHERE AND WHEN

To a few, the emergence of life on Earth has seemed so improbable that they have envisioned it as a legacy, or gift, from elsewhere in the universe. Lord Kelvin, whose views on the age of the Earth we encountered in an earlier chapter, argued fervently that meteors carried germs or spores from other worlds to our own. His motivation for proposing this idea, called *panspermia,* in 1871 was a desire to offer an alternative, more akin to Scripture, to the hypothesis that life evolved spontaneously. In one of Charles Darwin's less guarded moments (actually in a letter to a colleague), he spoke of the possibility that life had first emerged from organic matter in a warm, little pond. The implication, of spontaneous creation without a necessary role for the Creator, horrified Lord Kelvin. In contrast, panspermia required that God had imparted life's vital force to the living matter now piggybacked on meteors, even though this idea has been likened more to a holy sneeze from afar than a purposeful touch.

Neither religious leaders nor the scientific establishment readily accepted the idea of panspermia. In fact, it is probably fair to say that both camps were aghast at Lord Kelvin's bold proposal, though each for different reasons. Nevertheless, during the ensuing century a few like-minded scientists have reported the existence of biologically created molecules, living cells, and even fossilized organisms in meteorites. (In the reverse order—interestingly, the complexity of life reported from meteorites has steadily diminished, from the fossil remains of sometimes recognizable animals and plants, to tiny fossil microbes, and thence to chemicals that once were parts of organisms.) None of these reported discoveries has withstood scientific scrutiny, however. The latest version of panspermia, offered by Francis Crick, one of the discoverers of the structure of DNA (the cell's custodian of genetic information), is that the gift of life was carried to Earth on a spaceship sent by distant, extraterrestrial beings. There is not a shred of evidence for that claim. Panspermia, in whatever guise, is at the fringe of scientific credibility, and it is generally agreed that terrestrial life arose on its homeworld. Whether it did or did not is somewhat irrelevant, however; life arose somewhere, and we seek to understand the principles of the transmogrification that led to this momentous event. Arguing that life emerged on another world explains nothing; it merely removes the problem to a place so remote and unknown that we must abandon the quest to solve it. Since living forms use the common chemicals of Earth, it makes sense to proceed on the assumption that they arose here. Even though we may not be able to specify precisely how it happened, informed speculation seems preferable to the desperate view that life was imported from elsewhere in the universe.

We can say with certainty that life first appeared on Earth very long

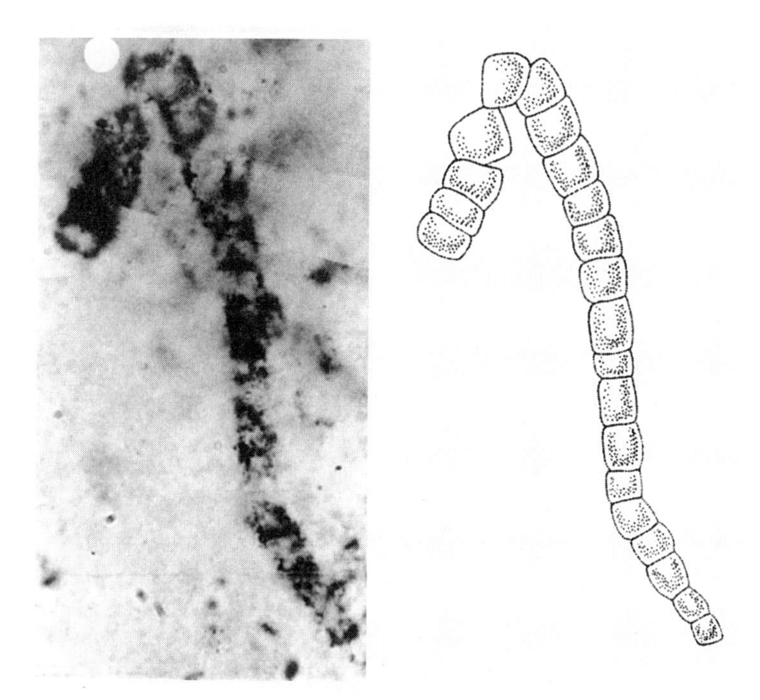

This filamentous bacterium was recovered from the 3.5-billion-year-old Warrawoona Formation in northwestern Australia. The individual cells are only a few micrometers in size. (Photograph and drawing by J. Schopf, University of California, Los Angeles)

ago. An initial, tenuous suggestion of the existence of organisms is found in sediments deposited 3.8 billion years ago. Deposits of carbon in these rocks are enriched in the light isotope carbon-12, relative to carbon-13, a hint that the carbon was once part of living organisms. Photosynthesizers tend to select carbon-12 in preference to the heavier isotope. The idea that these ancient carbon clumps are chemical fossils remains controversial, because the metamorphism that later affected these rocks may have modified the isotopic composition of their carbon. Not until hundreds of millions of years later does unequivocal evidence of organisms appear. In the Barberton Mountains of Swaziland (southern Africa) and at a few localities in Western Australia, sometimes referred to as the "North Pole" because they are so remote, are 3.5-billion-year-old sediments that contain the fossils. Actually, these tiny filaments, similar in appearance to modern bacteria, are more properly termed *microfossils*.

We owe the discovery of early microbes to Elso Barghoorn, a paleontologist at Harvard, and Stanley Tyler, a geologist at the University of Wisconsin. In 1952, Tyler sought Barghoorn's help in identifying some minute filaments in samples of ancient chert, and from that discussion sprang the recognition of the world's oldest life. Continuing the tradition, Barghoorn's former students, especially William Schopf and Andrew Knoll, have made careers of collecting the oldest sediments on Earth, painstakingly sifting through them for evidence of miniature life forms. Where most geologists might only see tiny, nondescript specks and strings, Barghoorn and a few others similarly infected with an appreciation for Archean microfossils found a buried treasure chest of antique organisms.

The microbial remains are associated with *stromatolites,* mounds of laminated rock that constitute the bulk of the Archean fossil record. These features also occur in the modern world, though they are uncommon. In shallow seas, mats of bacteria are periodically buried by thin films of sediment, impeding photosynthesis and thus killing the organisms. Some bacteria, by abandoning the dying cells and moving upward through the sediment, are able to form a new living layer. When this happens repeatedly, laminated mounds result.

The occurrence of stromatolites also leads to another point worth emphasizing: Life clearly originated in the oceans. In fact, nothing, neither plant nor animal, sprouted on or squirmed onto dry land until billions of years later, so far as we know. Even then, cells carried the saline ocean along within them, so that terrestrial creatures became ambling sacks of seawater. Although stromatolites formed in shallow water depths, current thinking is that mid-ocean ridges offer appealing sites for the first emergence of life. Clustered around these warm, nutrient-rich vents at the present day are bizarre organisms, including bacteria, tube worms, and blind shrimp, unlike those that live in the rest of the dark, frigid abyss. It is not hard to envision this as a nurturing environment for fledgling life forms. At the time of life's origin, the world was pelted more frequently with large meteors, and the ocean bottom may have been shielded from the effects of massive collisions. Life, especially organisms that sought the Sun in the ocean's photic zone, may have been exterminated, perhaps more than once, by such impacts.

The rather sparse fossil record of earliest life indicates the existence of nothing more complex than simple microbes and, judging from their morphology, life seems to have evolved at a snail's pace. Even more than a billion years after the first appearance of fossil remains, life remained surprisingly simple. For example, Proterozoic fossils in the Gunflint Iron Formation (1.9 billion years old) of Ontario, Canada, consist of bacteria similar in appearance to those that existed 1.6 billion years earlier. Mi-

Stromatolites, layered mounds of algae and sediment, in the Archean appear similar to those in modern environments. The upper photograph shows ancient stromatolites, from Victoria Island, Canada; below, a diver inspects modern stromatolites in the Bahamas. (Photographs by E. Shinn, U.S. Geological Survey)

crofossils from Spitsbergen, some 750 million years old, are more diverse but retain many similarities with their ancestors.

Although the fossil record suggests an Archean biota of simplicity and minimal change, a few words of caution are in order. Organisms of

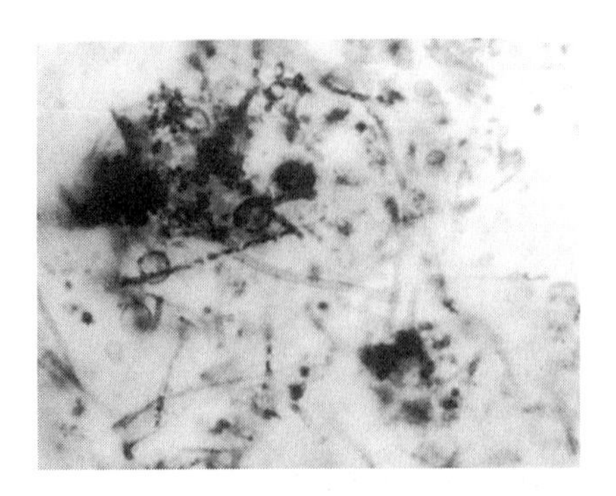

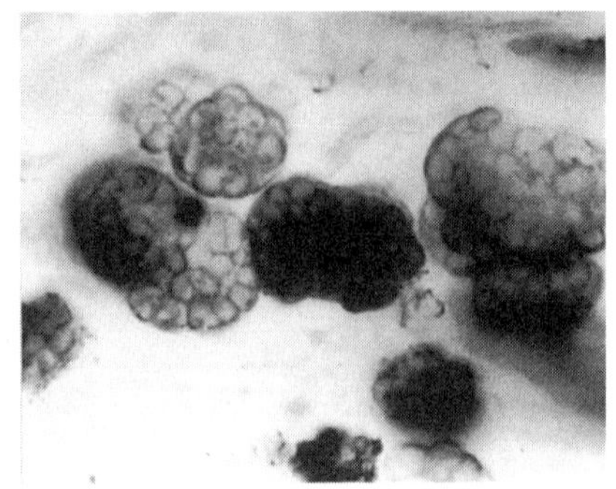

These microbes are representative of life forms occurring several billion years after the first appearance of cyanobacteria. On the left are tiny filaments and spheres of bacteria, from the 1.9-billion-year-old Gunflint Iron Formation, Ontario, Canada. The photograph on the right shows a more complex cyanobacteria in the Draken Formation, 750 million years old, from Spitsbergen. (Photographs by A. Knoll, Harvard University)

microscopic size are especially difficult to fossilize, and only those having tough gelatinous sheaths that maintain their shapes after the cellular matter within has decomposed are likely to be preserved. Consequently, most of the early fossil record consists of *cyanobacteria,* formerly known as blue-green algae, which have particularly resistant walls. There were probably other kinds of microbial slime populating the ancient oceans, but they disappeared without a trace. Because most outcroppings of Archean and Proterozoic rocks have suffered metamorphism that can destroy entombed fossils, we cannot really judge how widespread and diverse life may have been.

Even beautifully preserved microfossils carry only limited information. Fossilization is rarely rapid enough to preserve all of the parts of an organism before it decays, so what gets pickled in rock may be highly biased. All that can be discerned in microfossils, from our vantage point billions of years later, are the external morphologies of these tiny organisms. Like the Volkswagen, whose unchanging exterior over decades concealed many evolutionary changes in its internal machinery, many fossil microbes must have evolved without significant modification of their sheaths. As one example, the evolutionary transformation of *prokaryotes* (cells without nuclei, like the cyanobacteria) to *eukaryotes* (cells with nuclei) is difficult to pinpoint in the geologic record, because the nuclei are not usually preserved. Once the eukaryotes appeared, they spawned multicellular lineages that ultimately gave rise to fungi, plants, and animals; however, the history of the single-celled eukaryote ancestors is difficult to disentangle from that of the prokaryotes.

THE STUFF OF LIFE

It is no accident that *organic* and *organism* have the same root. The term *organic* originally described a branch of chemistry dealing with compounds produced by living organisms. These molecules invariably contain carbon, an element unique in its ability to form complicated linkages with other atoms. In 1828 the German chemist Friedrich Wöhler synthesized urea in his laboratory, thus demonstrating that life's vital force was not necessarily required in the making of organic compounds. Nevertheless, scientists for many years continued to associate organic chemistry with the peculiar reactions inside living organisms, a perception that was only dispelled by the discovery of organic molecules in outer space. Though the organic compounds found in interstellar clouds are relatively simple by the standards of organic chemistry, they testify to synthesis without benefit of biology. Moreover, the incorporation of interstellar organic matter into the materials of our solar system can be demonstrated by observations of comets, which release quantities of formaldehyde, acetylene, hydrogen cyanide, and other compounds as dust particles when they approach the Sun. Meteorites, too, contain organic matter, typically made of the elements carbon, hydrogen, oxygen, and nitrogen, and all having exotic isotopic compositions characteristic of interstellar matter.

I have to confess that the organic chemistry courses I took in college were not my favorite subjects. The field is mired in arcane terminology, requiring feats of memorization to recall the names and structures of huge molecules (not to mention the fact that organic laboratories often stink). There is a good justification for the emphasis on nomenclature, however. Carbon bonds with other elements, especially hydrogen, oxygen, nitrogen, and sometimes sulfur or phosphorus, producing extremely complex structures—straight chains, branching chains, pentagonal and hexagonal rings, and other bewildering forms. The ability to form linked networks is what makes carbon (as well as oxygen and nitrogen) so attractive for life; the multiple bonds yield strong but flexible structures suitable for constructing cell membranes and muscle fibers. The other elements that are commonly incorporated into living organisms, such as hydrogen, sulfur, and phosphorus, have properties that enable them to act as energy brokers or to perform other special functions.

From the myriad organic compounds that are possible, terrestrial life employs only a tiny handful. Proteins are the main structural and functional units of the cell. They are made from amino acids, each of which has the form

Hertzsprung-Russell Diagram

```
    H   H   O
    |   |   ||
H - N - C - C - OH
        |
        R
```

where H stands for a hydrogen atom, N for nitrogen, C for carbon, and O for oxygen. The NHH on the left side of the molecule is an amino group and the COOH on the right side is a carboxylic acid group, hence the name *amino acid*. R designates a side chain, normally consisting of some combination of carbon and hydrogen atoms. The various amino acids that make proteins differ only in the nature of the side chain. Terrestrial life employs only twenty amino acids from the vast array available to make proteins.

If amino acids are strung together end to end, like this

```
    H   H   O             H   H   O             H   H   O
    |   |   ||            |   |   ||            |   |   ||
H - N - C - C -(OH  H)- N - C - C -(OH  H)- N - C - C - OH   and so
        |                     |                     |           forth,
        R                     R                     R
```

and then the connecting water molecules (the HOH ovals can be rewritten as H_2O) are driven off, a small protein is produced. Real proteins are actually much longer than this example, and this abbreviated one is intended only to illustrate their relationship to amino acids.

Proteins have many functions, including the construction of cell materials, the consumption of energy, and regulation of the interactions of the cell with its environment. One thing they cannot do is replicate themselves. Within all modern organisms, the construction of proteins is actually guided by genetic codes housed in DNA and RNA (called *nucleic acids* because they may reside in the cell nuclei). Instructions for the order in which amino acids must be strung together to make specific proteins are encrypted in the DNA structure. DNA itself consists of long strands of four kinds of complex, ring-structured units (*nucleotides*), usually identified in shorthand by the initial letters of their constituent compounds: *A* (adenine), *C* (cytosine), *G* (guanine), and *T* (thyamine). The sequence of nucleotides in DNA, a kind of four-letter alphabet, specifies the order in which different amino acids should be attached in a growing protein. If this seems an improbable feat using so simple a template,

DNA Structure

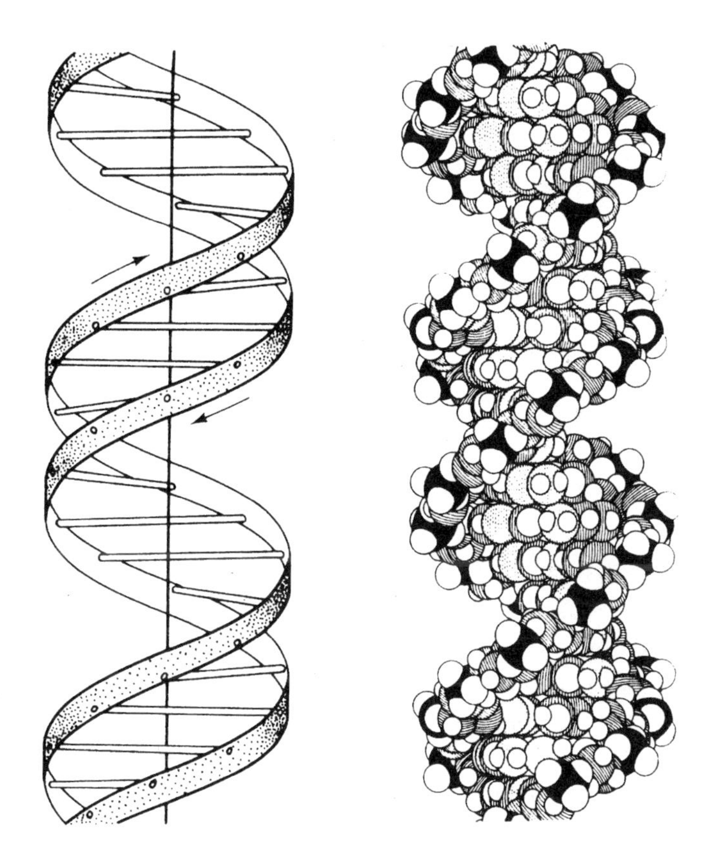

Two perspectives of the DNA double helix. The twisted side rails are made of sugar and phosphate (in the less schematic view on the right, white dots are hydrogen and carbon atoms and black dots are phosphorus atoms), and the cross ties in the center (dots with stippled pattern) are nucleotides.

remember that computers work their wizardry relying only on a two-letter alphabet, binary arithmetic.

DNA molecules in cells are actually formed by two intertwining strands of DNA that spiral around each other to form a double helix. The sugar-coated side rails of the DNA strands are made of sugar and phosphate molecules, and nucleotides form the cross ties. The opposing strands are bonded together in such a way that *A* is always paired with *T,* and *C* with *G,* like dancing partners. To replicate, the two strands unzip, and each then serves as a template to grow its lost complement. Opposite to every *A* in the template, a *T* is attached in the growing strand, and so on, as the devoted dance partners attract mates just like the ones they had before. Eventually a new DNA strand, identical to the one that separated, is formed.

Actually, DNA does not directly construct proteins, but instead performs that function through its lackey, RNA. We might say that DNA is an architect that always remains in its comfortable office (the cell nucleus), and RNA is the contractor hired to do the manual labor. The RNA molecule is a close relative of DNA and carries the same genetic information, except that a new nucleotide, *U* (uracil), replaces *T* in the sequence. The RNA strands carry the protein blueprint out into the cell and, using that, construct the protein molecules.

ALL FOR ONE, ONE FOR ALL

From a chemical perspective, all modern terrestrial life forms share uncanny similarities. Virtually all organisms contain the essential elements carbon, nitrogen, and phosphorus in the same ratio, 105:15:1, respectively. Over and over again, these elements, plus hydrogen and oxygen, are combined into proteins composed of the same twenty amino acids, DNA and RNA constructed of the same five nucleotides, and polysacharides formed from a few simple sugars. Bacteria, bison, and bananas all employ the same genetic coding system, and use the same energy-rich molecule (adenosine triphosphate, a mouthful of organic nomenclature thankfully abbreviated as ATP) for energy transfer in cells.

Another remarkable chemical commonality is the identical directions in which crystalline forms of biologic compounds rotate light. Carbon atoms can bond to four adjacent atoms or atomic groups to form distinct structures. If two otherwise identical molecules are oriented so that the atoms connected at the top and the bottom are the same, then those forming the "hands" can be interchanged, making left- or right-handed molecules. These complementary molecular configurations are called *stereoisomers*. Just as a screw with right-handed threads will appear to be threaded in the opposite direction when viewed in a mirror, two chemical isomers are mirror images of each other. They consist of exactly the same atoms in the same proportions, differing only in geometry. Crystals of stereoisomers twist polarized light either to the left or to the right, thus allowing the isomer to be classified as left- or right-handed.

Remarkably, the sugar molecules in DNA and RNA, from whatever biologic source, always rotate light to the right. Similarly, all of the amino acids from which proteins are formed, regardless of what plant

Stereoisomers

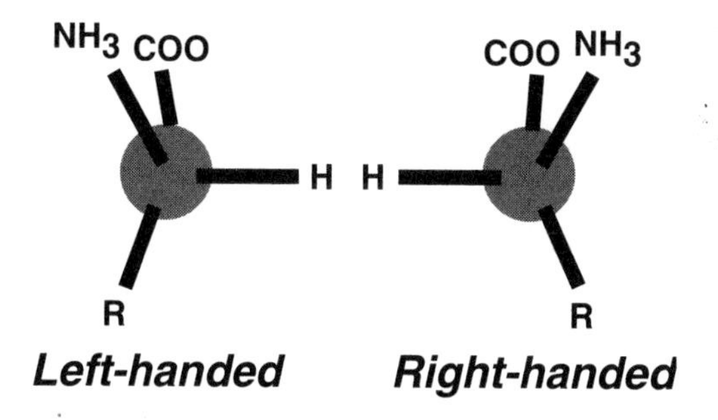

Stereoisomers are molecules that are mirror images of each other. Although both left-handed and right-handed amino acid isomers exist, all life forms use only the left-handed version.

or animal they come from, rotate light to the left. This peculiar situation appears to be necessary for life, because the cellular machinery that keeps organisms alive is constructed around the fact that genetic material always veers right and amino acids veer left. However, organic matter synthesized in a test tube or extracted from a meteorite does not have this property. Nonbiologic hydrocarbons consist of equal mixtures of left- and right-handed forms, so that light rotation to the left and right is canceled out.

What does all this mean? Why is the chemistry of all terrestrial life so uniform, and why do all organisms use the same stereoisomers? It probably points to the fact that everything alive evolved from some common ancestor, a microscopic Adam, an Urschleim of some description. The peculiar but similar chemistry of terrestrial life must have been inherited from this remote life form. As to why this common forebear should have had the peculiar chemical properties that it did, we can only speculate. The chance incorporation of certain chemicals and isomers may have dictated some functional advantage, or it may have been prompted by the environment. It seems possible that there must have been numerous, perhaps repeated attempts at early life, resulting in a variety of fledgling forms based on differing biochemistry. These organisms, incapable of making their own food, would necessarily have subsisted on other organic matter and, perhaps, each other. Our common ancestor may merely have been the one self-sustaining form that gobbled up its contemporaries. If several competing, biochemically distinct organisms had survived or had originated at later times, we should see that in the biochemical lineages of living organisms, but we do not.

All this is not to say that chemical differences between organisms do not exist, only that they are relatively minor, second-order effects. These small differences can be useful, however, in determining the degree of kinship and the rate of evolutionary change between organisms. Similar substitutions in RNA strands and in the amino acids of proteins may indicate that different organisms are more closely related than those that do not share these substitutions. In many cases, the chemistry (say, for example, the precise amino acid sequence in a particular protein) in a variety of plants or animals differs very little, presumably because any substitution would jeopardize the function of that protein. In a few cases, however, substitution has occurred and, more important, a consistent amount of substitution seems to have occurred over time. This situation allows the formulation of a molecular clock. For example, substitution of one type of amino acid for another in hemoglobin (a giant molecule with long chains of amino acids, used in the blood to carry oxygen) has occurred in animals that first appeared at different times in the fossil record. The rate of substitution is thought to have been more or less constant, so this may function as a molecular clock. Using it, the times of evolutionary divergence of other animals without good fossil records can be inferred from analysis of the chemistry of their blood.

Whether walking upright or slithering through the grass, whether sprouting from the soil or causing an infection, all of terrestrial life is closely related. The astounding truth of biochemistry is not just that humans and apes share a common descendant, but that we are also kin to the bumblebee, the oak tree, and the mushroom.

CHEMISTRY FIRST, BIOLOGY LATER

Where did these organic chemicals, used to form the dawn organism and passed down through innumerable generations, come from? There are several possibilities.

In 1924, Russian biologist Aleksandr Oparin proposed that life arose on Earth from an organic soup, a primordial slough of complex molecules cooked by lightning. He envisioned an acrid atmosphere choked with methane and ammonia, which reacted with seawater to yield life-giving chemicals. Nowadays, such a bouillon would quickly turn rancid, becoming food for microorganisms. In a world without hungry life, however, the organic products of such reactions could slowly accumulate and form a concentrated broth.

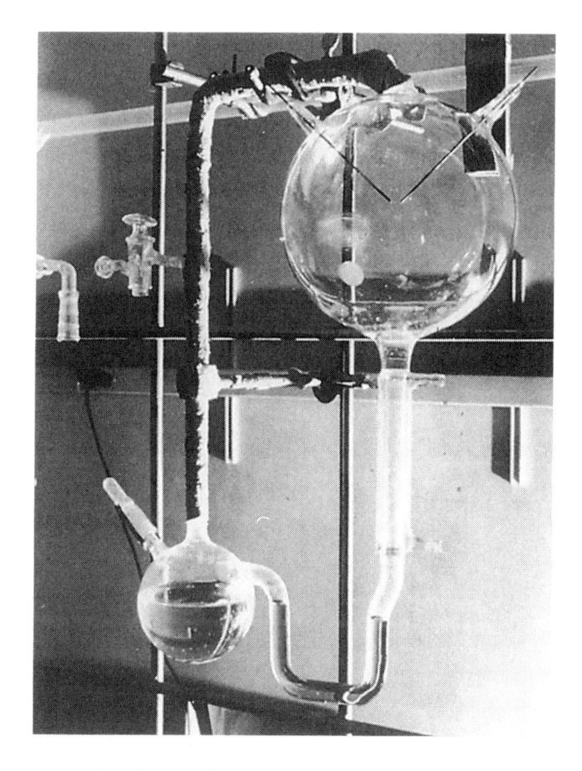

Using this apparatus, Stanley Miller simulated the formation of amino acids from simpler molecules during thunderstorms. In the upper flask he placed a primordial "atmosphere" of hydrogen, methane, ammonia, and water vapor. The electrodes inserted into this flask produced discharges, and the resulting organic matter collected in the flask at the lower left. (Photograph by S. Miller, University of California, San Diego)

Not until thirty years later was this idea experimentally tested. In 1953 Stanley Miller, then a graduate student at the University of Chicago, placed gaseous methane, ammonia, hydrogen, and water vapor into a flask. After repeated electrical discharges had flashed through this mixture for a few days, the gases turned crimson. Miller then condensed the contents of the vial and analyzed them. He had created amino acids galore. This result was immediately hailed as evidence that life's basic ingredients arose spontaneously in turgid pools on the early Earth. Subsequent experiments like this have produced even more complex molecules, including sugars and other components of nucleic acids.

This textbook example of the first hesitant step in life's origin is now under serious attack. To see why, we must revisit Oparin's notion that the atmosphere contained methane and ammonia. We now know that neither of these gases is stable for long in the presence of sunlight. Ultraviolet radiation rapidly breaks down these gases into carbon dioxide, nitrogen, and hydrogen. Consequently, many scientists now favor a carbon dioxide–rich atmosphere for the early Earth. The reason that the atmospheric composition is so important is that complex organic molecules are created much less efficiently from carbon dioxide and nitrogen than from methane and ammonia. It seems as if Miller's experiment was done under contrived conditions, although they seemed plausible at the time. Without the right atmospheric starting composition, it is doubtful that the synthesis of complex organic molecules could proceed.

A number of scientists still hold tenaciously to the idea that pockets of methane-and ammonia-rich atmosphere persisted long enough for organic synthesis to occur. Perhaps impacts with massive iron meteors might have altered the chemistry of air, albeit temporarily. All this is not really necessary, though, for there is another source of raw ingredients. Amino acids occur in chondritic meteorites, demonstrating that the formation of the same molecules used in organisms occurred in space. Of the seventy-four amino acids that have been identified so far, eight are present in proteins and eleven others have some biological role. Could the Earth have inherited its complex organic molecules from impacting meteors and comets, bypassing the first synthesis step in life's creation?

The tumultuous arrival of large objects and the ensuing cratering events are sufficiently energetic that most of the organic matter contained in planetesimals was probably vaporized. Tiny interplanetary dust particles, however, could have done the trick. These dust motes are slowed when they enter the atmosphere, and thus may have drifted down like snowflakes. Even today, countless particles, many containing organic compounds, rain out onto the Earth's surface. Chris Chyba, formerly a graduate student at Cornell University, has estimated that the collective mass of such particles is many thousands of times that of large meteors that strike the surface over the same time interval. At this rate of accumulation, organic matter equaling the mass of all present life could be dumped onto the Earth in only a few hundred thousand years. So, it seems plausible that the stuff of life may have been inherited directly from the cosmos. (I should emphasize that this is not panspermia revisited; the cosmic chemicals are not, and never were, alive.)

The dramatic import of this finding seems to have been overlooked, to judge from most science texts and articles. Accreting the organic building blocks of life from space actually solves a persistent riddle, the speed at which life seems to have arisen. The requirement that the Earth had to synthesize the raw materials for life from scratch is superfluous, although many biochemists continue to prefer Oparin's organic soup as their first course.

ASSEMBLY INSTRUCTIONS

Amino acids are used in the construction of proteins, an essential component of modern life forms. But a protein, by itself, is not alive, so its synthesis is still a giant step from life. The modern cell requires not only proteins, but also DNA and RNA to carry out its necessary catalytic and reproductive functions. Which of these essential components came first? In worrying about how life came to be, we encounter an obstruction that would seemingly bamboozle the most ardent champion of materialism: each of these molecules is dependent on the other two for either its function or its manufacture. DNA provides the blueprint, but it cannot perform as a catalyst and requires RNA to carry its instructions into the cell. On the other hand, proteins are excellent catalysts but cannot reproduce. It is now recognized that RNA is the most versatile of the three critical molecules, and its occurrence in all modern cells suggests that it is a very ancient concoction. A growing number of scientists now favor the creation of RNA as the next crucial step in the origin of life, a scenario known as "the RNA world."

I have previously pictured RNA as the hired help of DNA, a gopher that carries out the instructions of its superior. That is probably a very unfair description. RNA can perform all sorts of functions that DNA cannot, including acting as a catalyst, stimulating reactions between itself and other molecules. Perhaps such a talented molecule might have even spurred its own replication without DNA's assistance. Seen in this light, RNA may have been the ancestor of DNA, not its handmaiden. In the processes of making proteins from the available amino acids and of reproducing itself, RNA may have made the jump from the nonliving to the living, from Urschleim to algae.

But this begs the question. How was the first RNA created? This seminal event did not have the benefit of Stanley Miller's flask to concentrate the necessary ingredients together. Random organic synthesis, as any chemistry student knows, produces an appalling mess of diverse compounds, gooey garbage that would presumably obstruct life-giving reactions. Faced with this sobering knowledge, some scientists have suggested the need for a template on which to build the first RNA molecule. Certain crystals might have provided attachment sites for carbon, hydrogen, oxygen, and nitrogen atoms, so that they were arranged in a systematic manner and caused to react in a controlled way. The structure of clay, with its two-dimensional sheets and myriad surface-connecting points, might have served as scaffolding for the fabrication of organized molecules. Constructing life from a lump of clay seems to be an idea with scientific as well as biblical adherents.

A next, or perhaps concurrent, step was the formation of a membrane to compartmentalize the living organism. This may have occurred through the action of peculiar, two-faced molecules called *amphiphiles*. One side of these molecules has an affinity for water, and the other side is repelled by it. As a consequence, the molecules may have hidden their water-hating sides away by curling into tiny spheres. Their interiors would have been ideal sites for sequestering organic matter and carrying out chemical reactions by the earliest RNA molecules. We might view these delicate globes, bobbing in the ocean, as the ancestors of the first living cells.

OXYGEN POLLUTION

At the onset of life, Urschleim reveled in an environment free of elemental oxygen. Modern organisms have become so inured to this substance that they have forgotten that it is poison. Our cells remember, however. All oxygen-dependent, that is *aerobic,* organisms, human beings among them, must produce enzymes that detoxify oxygen. Although we cannot use this substance in the manufacture of other compounds, we depend critically on oxygen for the metabolism that provides our energy.

It has not always been so. Most of the early, prokaryotic microbes relied on the fermentation of organic molecules for their survival. Even after the advent of photosynthesis, most organisms derived their energy from smelly hydrogen sulfide and similar molecules. Such critters are still with us, hiding furtively in dank swamp mud, around hydrothermal vents on the ocean floor, and in our digestive tracts. One notable exception—no, that is too mild a term—one revolutionary lineage, the cyanobacteria, learned to use water in photosynthesis and, in the process, changed the world forever. These blue-green microbes became the most successful organisms on the planet, at least as can be judged from their production of biomass. Their waste product was free oxygen, a substance so toxic that it killed competing organisms, thereby conveying an edge to the cyanobacteria in the fight for survival.

The oxygen pollution inadvertently crafted by these tiny organisms changed the very composition of the atmosphere and set a new course for evolution. But this change was ponderously slow in coming. Cyanobacteria released oxygen into the atmosphere for more than 1 billion years before the effects first became apparent. By about 2 billion years ago, oxygen levels exceeded the critical threshold of a few percent of today's atmospheric concentration, enough to allow the appearance of aerobic organisms. Eukaryotes were possibly already present by this time, but the oxygen revolution allowed them to flourish. Some eukaryotes achieved prominence not because they learned to breathe themselves, but because they swallowed other bacteria that had learned the trick of respiration. Through this symbiosis some of these tiny but talented organisms became integral parts of the cells that had ingested them, in time evolving into energy-producing clumps called mitochondria.

Oxygen levels throughout the Archean and most of the Proterozoic eras were still not high enough to support our own lineage. Animals are conspicuously absent from the fossil record until much more recently, beginning about 600 million years ago. Nevertheless, it is the slow but inexorable rise of free oxygen, engendered by the microscopic cyanobacteria, that made this possible.

MICROBES ON MARS?

In 1984, an Antarctic expedition searching the Allan Hills for meteorites collected an unusual specimen. Unfortunately this meteorite, prosaically named ALH84OO1, was misclassified and remained largely ignored for a decade, but it was finally revealed to be a member of a small group of meteorites thought to be derived from Mars. It is very old, unlike other Martian meteorites, and was probably derived from the ancient, heavily cratered southern highlands. This particular stone is heavily fractured, and small beads of carbonate occur within the void spaces. Almost overnight, these tiny carbonate grains have become arguably the most interesting matter on Earth. They may contain evidence for extraterrestrial life, perhaps something akin to the earliest life on Earth.

The recent suggestion that a meteorite from Mars contains microbial remains of fossilized organisms has regenerated public interest in the idea that life is abundant, at least on worlds that contain (or once contained) water. As intriguing as are the observations of the carbonates by David McKay of the NASA Johnson Space Center and a number of colleagues from various institutions, however, they are not yet sufficient proof for a hopeful but largely skeptical scientific community. As expressed by one scientist at the NASA press conference where the observations were unveiled, extraordinary claims require extraordinary evidence.

McKay and his colleagues analyzed organic compounds characterized by rings of carbon (polycyclic aromatic hydrocarbons, or "PAHs") associated with the carbonates, and they interpreted the PAHs as the products of decayed organisms. PAHs also form in smoke stacks, candle flames, and automobile mufflers without any help from biology, and they are common constituents of many kinds of extraterrestrial matter. Unfortunately, it is impossible to tell whether the assortment of PAHs in the Martian meteorite formed by biologic or inorganic processes. Another observation cited as evidence for life was the occurrence of miniscule grains of iron oxide (magnetite) and iron sulfide (pyrrhotite) within the carbonate beads. Although it is true that these minerals resemble those formed by some bacteria on the Earth, most crystals in this tiny size range have similar appearances and their size and shape may not be diagnositic of a biologic origin. Terrestrial bacteria string magnetite grains together into beads, which allow them to sense the Earth's magnetic field and orient themselves within it; magnetite chains have not been reported in the Martian meteorite. Most interesting are the tiny elongated forms suggested to be fossilized nanobacteria (the prefix "nano" refers to their extremely small sizes, measured in nanometers, or billionths of meters). There is some controversy among biologists concerning exactly what nanobacteria are, and even whether or not they are alive. In any case, the morphologies so far described in the Martian meteorite do not appear to be convincing microfossils, without images of cell membranes, cells in the process of division, or other diagnostic observations.

It is entirely plausible that Mars once harbored primitive life forms, and their discovery would be a major scientific triumph. The Martian environment prior to about 3.5 billion years ago is thought to have been warmer and wetter than at present, because sinuous river channels are carved into the ancient highlands and the geologic formations there show evidence of higher erosion rates. Perhaps simple life forms once lived in Martian groundwaters or in saline lakes that temporarily filled impact craters. If so, their very existence could imply that life arises wherever and whenever conditions permit. Such microorganisms might even be alive today, though it seems likely that any extant life must have retreated, along with water, into the subsurface. If terrestrial experience is any guide though, fossils will be hard to find. The jury is still out in the case of ancient life in Martian meteorite ALH84001, and we will have to wait for further research to establish whether the curious features described in this stone really represent the vestiges of extraterrestrial microbes.

FREAK OR NATURAL CONSEQUENCE?

So what have we learned from this trek into the genesis of life? Everything now alive on Earth probably shares one primitive ancestor. This dawn organism arose from an organic brew billions of years ago, improbably assembled from the stuff of the cosmos and somehow nurtured to life by our own world's peculiar environment. From this earliest ancestor we inherited the rudiments of our biochemistry—the distinctive mixture of carbon, nitrogen, and phosphorus common to all life, the unique coupling of protein and nucleic acid that allows both function and replication, the handed geometric forms that twist light in only one direction. Over time living organisms emerged from the stench of hydrogen sulfide and the salty slosh of the ocean, but even we highly evolved land dwellers cannot evict the remembrance of past homes from our cells.

Regardless of the precise chemical and geological processes and pathways that led to life, it may be that its appearance was no freak accident. As expressed (somewhat clinically) by a prominent biologist and Nobel laureate:

> . . . these [*life*] processes were inevitable under the conditions that existed on the prebiotic Earth. Furthermore, these processes are bound to occur similarly wherever and whenever similar conditions obtain. This must be so because the processes are chemical and are therefore ruled by the deterministic laws that govern chemical reactions and make them reproducible.
>
> —Christian de Duve, *American Scientist*

It is true that the rigid rules of chemistry require that reactions between atoms or molecules happen whenever and wherever they are energetically favorable. It is also true that nature exhibits a pronounced tendency to self-organize, suggesting that the eventual appearance of life may have been more deterministic than miraculous. A strict materialist view, as expressed in the foregoing passage, regards life processes as the manifestations of chemical reactions that, by their very nature, should have been unavoidable under the right conditions.

As a scientist, I readily acknowledge the validity of this logic. But is this explanation for life, at least life as we know it now, sufficient? Do biochemists have the last word, indeed the only word, in describing life's breathtakingly clever capabilities as the inevitable responses of combining chemical reagents? My daughter once had a rabbit (the gray, floppy-eared variety), memorable from among all her other pets to me only because I watched it die in her arms. At the instant of death, did the mad rush of chemicals from this creature back into the Earth signal only the shutting down of its internal chemical factory, a straightforward severing of molecular interactions, or were my daughter's tears the proper response for the loss of something more intangible? I do not know. I have to confess that, faced with the elegance of life's machinery and the intricacy of its consequences, it is sometimes difficult to accept chemistry as quite enough. To quote another who has thought about these things more than I have:

> Men talk much of matter and energy, and of the struggle for existence that molds the shape of life. These things exist, it is true; but more delicate, elusive, quicker than the fins in water, is that mysterious principle known as "organization," which leaves all other mysteries concerned with life stale and insignificant by comparison. For that without organization life does not persist is obvious. Yet this organization itself is not strictly the product of life, nor of selection.
>
> —Loren Eiseley, *The Immense Journey*

This leads to an interesting philosophical point: Even if science succeeds in its steady descent down the ladder of time to the very point of life's inception, even if living molecules can somehow be constructed in a sterile laboratory, will we finally understand what it means to be alive? At some point we might say, with some assurance, that life arose from a particular ancestral nucleic acid, which organized itself into the first cell by way of this particular reaction and that. But, as Eiseley further notes in *The Immense Journey,* that alone is not the answer to "the saw-toothed grasshopper's leg," nor to "the subtle essences of memory, delight, and wistfulness moving among the thin wires of my brain." I mean no disrespect to my fellow scientists who work among outcroppings and test tubes with laudible patience and ingenuity, seeking the origin of life. But I cannot help but wonder if fiddling crickets and pondering humans may have become more than the sum of their chemistry.

Some Suggested Readings

Cairns-Smith, A. G. (1985). *Seven Clues to the Origin of Life.* Cambridge University Press, New York. *This engrossing little detective story makes the case for crystals as templates for life; it is fascinating reading, although it goes one step too far for my taste.*

Cloud, P. S. (1988). *Oasis in Space: Earth History from the Beginning.* Norton, New York. *A lively and interesting introduction to the beginning of life, by an eminent authority in the field.*

de Duve, Christian (1995). *Vital Dust: Life as a Cosmic Imperative.* Basic Books, New York. *This up-to-date book describes, in rich and easily readable detail, the chemistry of life and ideas about its origin.*

Eiseley, Loren (1957). *The Immense Journey.* Vintage Books, New York. *The musings of a thoughtful scientist and elegant writer from the previous generation: I especially recommend the last chapter ("The Secret of Life") for a different perspective on life's origin.*

McSween, H. Y., Jr. (1993). *Stardust to Planets: A Geological Tour of the Solar System.* St. Martin's, New York. *The final chapter of this book describes the history of panspermia and the possible role of impacts in the origin of life more fully than in the present book.*

Orgel, L. E. (1994). The origin of life on the Earth. *Scientific American,* vol. 271, no. 4, pp. 77–83. *The RNA world—a world in which RNA promoted the reactions necessary for life's common ancestor to survive and replicate—is discussed in this interesting article.*

Oro, J., Miller S. L., and Lazcatto, A. (1990). The origin and early evolution of life on Earth. *Annual Reviews of Earth and Planetary Sciences,* vol. 18, pp. 317–56. *This is a highy technical description of attempts to unravel the chemistry of life; an excellent summary, but not for beginners.*

Sagan, Carl (1994). The search for extraterrestrial life. *Scientific American,* vol. 271, no. 4, pp. 93–99. *Although, as far as we know, Earthlings exist alone in the solar system, the discovery of extraterrestrial life would be profound. In this article, Sagan describes recent and ongoing attempts to determine whether life might have arisen elsewhere.*

Schopf, J. W., ed. (1992). *Major Events in the History of Life.* Jones and Bartlett, Boston. *An up-to-date survey of important topics in the evolution of life; especially relevant to the origin of life are chapters by S. L. Miller on "The Prebiotic Synthesis of Organic Compounds as a Step Toward the Origin of Life" and J. W. Schopf on "The Oldest Fossils and What They Mean."*

Monkey Business at the Seaside

The Evolution of Evolution

LIVING ON THE EDGE

Without some respite at the sunny, sea-beaten edges of the world, summers for me would be sadly incomplete. My family regularly vacations on barrier islands off the South Carolina coast. This string of overgrown sandbars, stretching like a pearl necklace along the Atlantic seaboard, is unlike most other beaches. Each island is separated from the mainland by a tidal marsh, so that the character of its backside is markedly different from its wave-swept front. The marshes hugging these Janus-faced islands literally squirm with life of all description, revealed twice daily as the tides come and go. The flooded expanse of knife-edged spartina grass transforms at low tide into meandering creeks and stranded pools brimming with camouflaged flounder, blue crabs, and translucent shrimp. Receding water exposes banks of putrid mud so pocked with tiny fiddler crab burrows that they resemble Swiss cheese. And egrets with question-mark necks stand patiently amid shoals of oyster shells, dining at leisure on schools of flashing minnows. The sights, smells, and sounds of the salt marsh testify to the fact that this is the richest concentration of life on the planet.

The sea's edge (the marsh plus the nearby ocean shallows) always has been a veritable laboratory for biologic diversity. How did this amazing assortment of organisms arise? Certainly, a mullet would seem to have little in common with marsh grass, or a clam with a pelican. In the previous chapter, we focused on the origin of life and, in the process, learned that all of it derives from a common ancestor. Here, we will consider the evolutionary processes that have allowed life to erupt into riotous variety.

Great stacks of erudite books have been written on evolution, so I cannot hope to do it justice in one chapter. Instead, I will try only to illustrate some rudiments of modern evolutionary theory by considering a few instructive examples (and a few surprises) from the fossil record. It is appropriate that we begin this discussion at the seaside. There, many hundreds of millions of years ago, biology became synonymous with variety. And it was there, only a century and a half ago, that Charles Darwin first sat down to compose his blockbuster, *Origin of Species*.

A RELUCTANT ICONOCLAST

I don't know if Darwin shared my fondness for the seashore. True, he spent years on a round-the-world cruise exploring the shorelines of several continents, but extreme bouts of seasickness must have soured the experience. Because Darwin was a hoarder and never discarded any records, the details of his churning stomach and, more important, his scientific discoveries as naturalist on the H.M.S. *Beagle* are well documented.

Nothing escaped Darwin's attention. Sailing down the South American coast, he noted the strange animals that were so well adapted to their harsh environments. He also wondered about great fossil bones protruding from rocks and shrewdly tried to relate them to living animals. Sailing north, the *Beagle* encountered the Galapagos Islands, partly drowned volcanoes populated by armored tortoises, giant lizards, and finches in fantastic profusion. The last, especially, puzzled and captivated Darwin. Their beaks—of every shape and size, seemingly tailored for every conceivable purpose—existed in varieties unknown anywhere else. These caused him to fancy that "from an original paucity of birds in this archipelago, one species had been taken and modified for different ends." Before Darwin, the Galapagos fauna (or, for that matter, any island's inhabitants) were assumed merely to be the vestige of some mainland population, marooned when a prior connection to the nearest continent was submerged. His genius was in recognizing the true significance of this world in miniature, this isolated place where the forces at work to create new organisms could be plainly observed.

Darwin returned home to an England beset by anarchy, or so it must have seemed to him. Rioting socialists crowded the streets, shouting for

Charles Darwin, in caricature and as icon (with his famous finches). The British postage stamp was issued in 1982, along with several others depicting the Galapagos expedition, on the centenary of Darwin's death. (Courtesy of R. Milner, American Museum of Natural History)

revolution against an economic system that thrived on exploitation and workhouses. The subversives also railed against the institutions that encouraged this system, namely the state and the church. Arrayed against this rabble stood a conservative gentry who, if they could not maintain order over the dissatisfied masses, were at least comforted by a scientific establishment that comported with their religious beliefs. Darwin himself came from this conservative world of wealth and privilege and, prior to the voyage, he had even entertained the possibility of becoming a clergyman. He was an unlikely champion for a radical scientific theory.

For many years after his return, Darwin wrestled with his ideas about evolution and with their upsetting implications for mind, morality, and religion, castigating himself as the "devil's chaplain." He sought sanctuary in rural Kent, mostly avoiding his neighbors, and confined his thoughts to secret notebooks. The effort to reconcile evolution with his Victorian beliefs apparently left him perpetually ill. We now recognize the possibility that his migraines and nausea were psychosomatic, brought about by anxiety over his incendiary thoughts, although he may also have been infected with a parasite on his voyage. Whatever the cause of his illness, Darwin's fear of having to face a storm of personal criticism, and of the rending of the social fabric that his ideas might engender, explains why he procrastinated for two decades before publishing.

Against his wishes, Darwin was finally forced into publicly exposing his theory. A young, relatively unknown naturalist, Alfred Wallace, while on his own cruise to Indonesia, independently worked out the scheme by which evolution worked its magic. Ironically, it was to Darwin himself that Wallace, in innocence, sent his manuscript for review. Distraught at the thought of being 'scooped,' Darwin turned for advice to several scientific colleagues. These intimates, aware of the decades of painstaking research that might now go unrecognized, quickly intervened. They hurriedly arranged for a short summary of Darwin's work to accompany Wallace's paper. The theory that was to rock the world was thus launched simultaneously by two scientists.

When first read at a meeting of the Linnean Society, Darwin's paper was met with stunned silence. Some listeners certainly grasped the enormity of what they had just heard, and perhaps others were shocked at its audacity. Not everyone, though, was impressed. The Society president walked out, muttering that the year had not been marked by any striking discoveries. At that moment Darwin could not have cared less. He was burying his son, a victim of an outbreak of scarlet fever, in a parish graveyard, and the next day he would evacuate his family to another town.

Several weeks later, Darwin took his children and his nauseous stomach to Sandhurst on the Isle of Wight, to recuperate in the bright sunshine and the sea air. Now that he had finally been forced into the open, he recognized that he had to explain and document his theory more fully. Although still in mourning for his son, Darwin began to work on a book. He was pressed to move quickly, and *Origin of Species* actually turned out to be considerably shorter than the opus he had originally planned. Any fears he might have had that his life's work would be ignored shortly proved to be unfounded. When published in 1859, the first edition sold out in a single day. Darwin's fame was instantaneous, though "infamy" might better describe his reception in certain quarters. Wallace generously referred to the new theory as "Darwinism," minimizing his own relative contribution as but "one week to twenty years."

As everyone knows, Darwin's revelation was that evolution occurred by *natural selection.* This was, and remains, a relatively simple idea, consisting, as one prominent naturalist has noted, of two undeniable facts and one inescapable conclusion. The facts are as follows: (1) Organisms vary, and (2) some of that variation is inherited by their offspring, a population too numerous for all of them to survive. Darwin's conclusion was that the offspring best suited to their environment will be more likely to survive. Consequently, the better adapted tend to increase in number and, in the process, to steer life in the direction of those so selected.

An important part of Darwin's theory was that variation arises without pattern, continually generating organisms that are randomly better or worse fitted to their environment. It follows, then, that evolution has no guiding purpose, other than to promote the survival of the best suited and to ensure that there will be a supply of individuals that can respond to changing environments. This is the part that so troubled Darwin, a concept that many people still find abhorrent today. This dangerous idea undermined the fundamental assumption and fervent hope of human uniqueness, and of its special creation. Although *Origin of Species* scarcely addressed the question of human evolution (Darwin did that years later in *The Descent of Man*), it was inevitable that this aspect of his work would erupt into a firestorm of controversy. No matter their acceptance of evolution, friends and foes alike deplored this stumbling, haphazard mechanism. In time, however, science would embrace Darwin's ideas, a revolution that shook its very foundations.

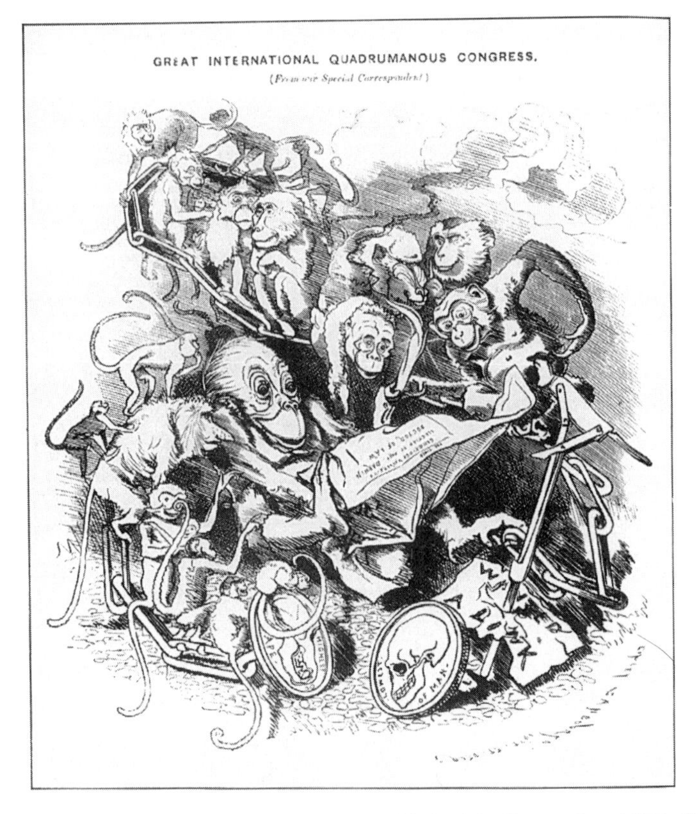

This *Punch* cartoon of the "missing link," published in December 1877, illustrates the difficulty many people had with accepting Darwin's theory of evolution by natural selection. The drawing was inspired by Cambridge University students, who lowered a monkey marionette from the gallery when Darwin was awarded an honorary doctorate.

THE DIFFERENCE BETWEEN FACT AND THEORY

Let me be clear about something here. Science views evolution as *fact,* not theory. The question is not whether evolution happens, but how it happens. Natural selection is Darwin's theory about the means by which evolution occurs. Scientists sometimes argue about the details, such as the precise nature of the copying process or the speed with which changes occur. With very few exceptions they do not, however, dispute the fact of evolution.

Darwin did not think that you could watch evolution in action. The process, he wrote in *Origin of Species,* works "silently and insensibly." But in this he was wrong. Modern science has rendered evolution visible by using computers to track physical changes from generation to generation. Furthermore, molecular biologists can now follow some of these changes at the genetic level. Many examples of evolution are familiar. Breeders of plants and animals know that, by carefully selecting their stock, they can change the characteristics of their crops or herds. The annual appearance of new strains of flu virus testifies to continual biologic change without the intervention of overseers. Even Darwin's famous Galapagos finches have evolved slightly different beak shapes since his visit, a response to changes in the sizes of available seeds. A particularly interesting example relates to the ominous spread of infectious diseases. That evolutionary change is prompted by the environment can be seen in the appearance of bacteria that are resistant to an antibiotic-drenched world. Drugs that kept these infectious agents at bay for the past fifty years now no longer destroy them. The indiscriminate use of antibiotics has selected for increasingly resistant bacteria, an unpredicted experiment in evolution by Darwin's favored mechanism.

Admittedly, most of these modern examples involve the modification of characteristics within species rather than the appearance of entirely new species, but the principle is the same. The appearance of new species must happen in the modern world, but it is not possible to distinguish a newly evolved species from the discovery of a previously unknown one that has been around for some time. Even though Darwin had no notion of how the characteristics were transmitted (that is, through the genetic code in DNA), the fundamentals of our understanding of evolution by natural selection remain more or less as he originally specified.

The most instructive examples of evolution lie entombed in the geologic record, where enough time has elapsed for evolution to work wonders. Before we examine a few facets of this great unfolding, some perspective is in order. There is a natural tendency to view evolution in sequence, as progressively higher forms supplant lower forms in the geologic record. A more accurate reading, though, shows that many organisms have missed the bandwagon, and it would be a mistake not to recognize that fact.

> This is not the "age of man"; it is not even the "age of insects"—a proper designation if we wish to honor multicellular animal life. As it was in the beginning, is now, and ever shall be until the sun explodes, this is the "age of bacteria." Bacteria began the story 3.5 billion years ago, as life arose near the lower limits of its preservable complexity. The bacterial mode has never altered; the most common and successful forms of life have been

> constant. Bacteria span a broader range of biochemistries and live in a wider range of environments; they cannot be nuked into oblivion; they overwhelm all else in frequency and variety; the number of E. coli cells in the gut of any human exceeds the count of all humans that have lived since our African dawn.
>
> —Stephen Jay Gould, *Natural History*

But let us not be awed by biomass. The importance of the evolution of higher forms of life does not lie in their volume, but in their complexity. This is not something, by the way, that is inherent in Darwin's theory. Natural selection is biased, but it does not necessarily lead to the evolution of higher life forms; it just did in many cases, a result for which we all should be thankful. Evidence from the geologic record affirms Darwin's guiding mechanism of evolution but, as we will shortly see, it also reveals that there is more to the story.

LESSONS OF THE CAMBRIAN EXPLOSION

Darwin was unflaggingly honest in exposing weaknesses in his theory. An apparent explosion of diverse life forms at the onset of the Cambrian period headed the list. (The Cambrian is conventionally thought to have begun about 570 million years ago, although some geologists have suggested a younger age.) In fact, Darwin devoted an entire section of *Origin of Species* to this vexing subject. His theory of gradual evolution by natural selection required that Cambrian organisms must have had ancestors, but to his dismay, not a single fossil had been found in older strata. It was as if the many species of the Cambrian had suddenly appeared out of thin air, an inference understandably relished by creationists. Chief among these was Roderick Murchison, a prominent British geologist who played an instrumental role in defining the record of early life. Murchison considered the Cambrian explosion as strong evidence against evolution. In fact, he argued that it was God's moment of creation.

As his response, Darwin fell back on a standard argument to explain missing ancestors: that the fossil record was fragmentary and incomplete. We now recognize that this argument was valid, at least to a point. Life in the Cambrian, for the first time, employed hard parts (shells, bones, and teeth), which were much easier to preserve than soft, mushy tissues. Indeed, most of the fossil record from the Cambrian onward consists of skeletal parts, which sometimes poses a problem to paleontologists' ability to reconstruct an organism's anatomy from its remains. Documentation of life before the Cambrian had to wait for a hundred years, until geologists finally discovered casts of soft-bodied animals and fossil microbes.

Cambrian fossils are common almost everywhere there are marine sedimentary rocks of this age. Nowhere are they preserved so exquisitely, however, as in the Burgess Shale of British Columbia, Canada. Because they were quickly buried by mud slides in the shallows of the Cambrian sea, their soft parts were preserved and so their anatomies can be reconstructed in great detail. Paleontologists Harry Whittington of Cambridge University and his former students Derek Briggs and Simon Conway Morris have invested decades in meticulously unraveling the secrets of the Burgess Shale, and what has emerged is surprising. This formation contains 120 genera, including trilobites, brachiopods, echinoderms, cnidarians, sponges, worms, and the earliest known chordate. The most remarkable story, however, lies in the creatures that do not fit conveniently into any of the categories above. The Burgess fauna contains a range of anatomical design that has never again been equaled, possibly more variation than exists now in all the world's oceans. To introduce but a few of the many bewildering organisms in this bestiary: There was *Marrella,* the "lace crab," a skittering arthropod with multiple spines, numerous body segments, and delicate gills. The armored sluglike creature *Wiwaxia* protruded protective spines as it crawed along the seafloor, periodically molting to leave behind a vacated husk. Predatory *Anomalocaris,* two feet in length and bearing grasping appendages, ate with a circular mouth that functioned like a nutcracker. This nightmare has been implicated as the cause of semicircular bites on the sides of Cambrian trilobites. And, perhaps oddest of all, the aptly named *Hallucigenia,* whose appearance defies our ability even to specify which end is the head, or which side is up.

All of these weird wonders represent extreme variations in the design of life, pioneering patterns that are no more. In fact, some of the Burgess creatures cannot be placed confidently into our existing classification scheme for living creatures. They apparently lie outside the bounds of modern phyla; they were not the ancestors of any modern organisms. Mother Nature experimented with all kinds of anatomical designs in the Cambrian, many of which were found wanting for one reason or another.

Darwin was right to worry that the Cambrian fauna posed a problem

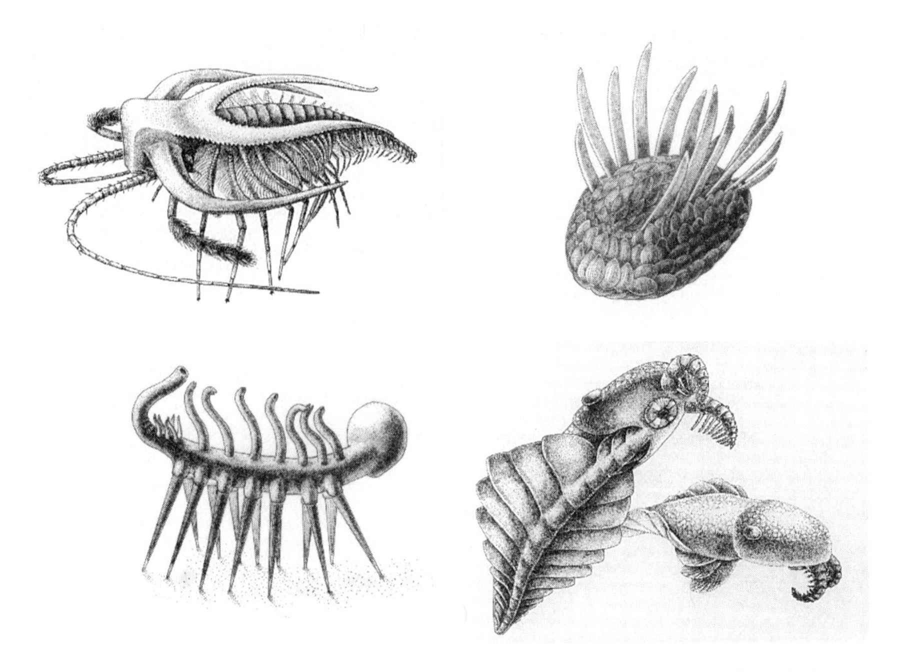

These elegant reconstructions of Burgess Shale animals were drawn by Marianne Collins. Beginning at the upper left and moving clockwise, they are of *Marrella splendens* (the tiny "lace crab"), *Wiwaxia* (an armored slug), two views of *Anomalocaris* (the carnivorous terror of the Cambrian), and *Hallucigenia* (a bizarre creature with an apt name). (From *Wonderful Life: The Burgess Shale and the Nature of History* by S. J. Gould; reprinted by permission of W. W. Norton & Company, Inc.)

for his theory of evolution by natural selection, but the problem was not the absence of forebears. Instead, it lies in the cause of their demise. The Burgess Shale fossils employ many basic body plans, only a few of which survived and blossomed into successful groups. Traditionally, the Cambrian has been viewed as a time of experimentation, with anatomical suitability determining which species survived and which became evolutionary dead-ends. There is a danger of circular reasoning in advocating natural selection, as recognized by Darwin himself. Survival is the phenomenon to be explained by superior adaptation, not the proof of it. If fitness is defined by survival, then "survival of the fittest" degenerates into the meaningless "survival of the survivors." The solution to this quandary is that survival must be *predictable,* based on superior form and function for a certain environment. In the case of the Cambrian fauna, detailed studies have not yet been able to establish that those few organisms that survived the Cambrian possessed functional advantages over those that did not. The beast *Anomalocaris,* for example, disappeared, despite its well-honed ability to eat its neighbors. Experts on the Burgess Shale have suggested that the choice of winners and losers cannot be predicted from inferences about their environmental fitness. Perhaps, in this instance, natural selection played second fiddle to luck. During the Cambrian, and at other times as well, some evolutionary changes may have been a random stroll through the possibilities.

Darwin, and most evolutionists since his day, assumed that the evolution of new species was a gradual process. But the fossil record, as exemplified by the Burgess Shale, tells a different story. It often seems that species appeared suddenly, without traceable lineages of transitional forms. This worrisome fact was addressed by paleontologists Niles Eldredge of the American Museum of Natural History and Stephen Gould of Harvard in the 1970s. Why not, they argued, accept the geologic record at face value? Perhaps new species really do appear suddenly, after long periods of stasis. Some Darwinists dismissed this idea as "evolution by jerks," but its authors insisted that the unseen evolutionary processes were guided by natural selection. In the ensuing controversy, some skeptical scientists sought to prove or disprove this idea by measuring the anatomical details of extensively sampled fossil organisms. Alas, they usually found yawning stretches with minimal change, punctuated by short spurts of morphological diversity. This concept, known as *punctuated equilibrium* ("punk eek" to its devotees), has now been incorporated by many evolutionists into Darwin's theory.

ONTO THE SHORE

Just before Christmas in 1938, Marjorie Courtenay-Latimer was puttering around the docks in East London, South Africa, looking for fish. These fish were not to be her supper; as part of her job as ichthyology curator at the local museum, she regularly searched for unusual specimens. On this day, Courtenay-Latimer got more than she bargained for. A trawler had snagged a strange-looking prize, a five-foot-long, pale blue fish with a tufted tail and muscular lobes anchoring its fins. The fish had already begun to decay, so she had to cajole a taxi driver to carry her and her smelly find to the museum. Within a few days, an associate had identified the fish as a *coelacanth* (pronounced "SEEL-uh-kanth"), thought to have been extinct for 50 million years. Unfortunately, by this time the fish had been gutted, so little scientific information could be gained.

This rare photograph of a *coelacanth,* the only living crossopterygian, was taken from a submersible. (Photograph by H. Fricke, Max Planck Institute, Germany)

When news of the discovery was circulated, it set off an immediate search for other coelacanth specimens. Not until fourteen years later was another finally dragged from the deep, this time near the Comoro Islands, far away from South Africa. That location proved to be fertile ground, so to speak, and since then more than a hundred specimens have been caught at that site. Living coelacanths have even been observed and photographed in their natural habitat. They are slow-moving fish, using their fins as a four-legged land animal walks—with pectoral and pelvic fins moving in opposite directions.

Coelacanths are the only living survivors of a once large and diverse group known as *crossopterygians,* the tassel-tailed fish. Crossopterygians of Devonian age (about 400 million years ago) are the closest known relatives and probable ancestors of the first amphibian landlubbers. How could this fish have made the trip to dry land?

Doubtless, it was an uncomfortable migration, and not at all stately.

> There are two ways to seek the doorway: in the swamps of the inland waterways and along the tide flats of the estuaries where rivers come to the sea. By those two pathways life came ashore. It was not the magnificent march through the breakers and up the cliffs that we fondly imagine. It was a stealthy advance made in suffocation and terror, amidst the leaching bite of chemical discomfort.
>
> —Loren Eiseley, *The Immense Journey*

The fossilized pectoral fin and shoulder bone of a crossopterygian is similar to the leg of the earliest land-dwelling amphibians. (Photograph by J. A. Long)

By some means unknown, this fish was prompted to squirm out of the marsh, at least for short periods of time, to face a much more hostile environment than the one it left behind. Nevertheless, the crossopterygian, more than any other animal, was prepared to take this formidable step.

If there are miracles in evolution, a prime example probably lies in life's adaptation to the land. Unfortunately, the fossil record provides only limited insights into how organisms coped with the many obstacles they faced. First and foremost, there was gravity. In the water, of course, gravity is no big deal. By balancing its buoyancy, a fish can effectively weigh nothing, so support and locomotion are easier. Life on land, however, requires a sturdy internal skeleton to counteract the force of gravity. The fossil record of crossopterygians indicates increased ossification of their skeletons, as well as the development of strong bones that would, in time, support shoulders and hips. Their fins already had the same bone pattern that all land animals would retain in their limbs, and the muscular knobs that controlled the fins may have provided the power for an able, though slow and ungainly waddle across the ground.

Eating on land posed a difficult problem, one that crossopterygians almost certainly did not bother to solve. Their anatomy was much better suited to feeding in the water, and there wasn't much to eat on land anyway. As evolution brought prey out of the water, however, their descendants had to develop the ability to make fast lunges.

Breathing atmospheric oxygen posed another hurdle. Fish take oxygen from water that is passed over their gills, and the pumping system that moves water through the mouth and out the gill openings could circulate air as well as water. But a new organ, a lung, was required to process air efficiently. The crossopterygians, like many modern fish, had swim bladders to regulate gases and thus control their buoyancy. This organ probably functioned as a primitive lung for the first hardy land explorers, as it does for the modern-day mudskipper, a popeyed fish that climbs trees with its fins and pursues insects.

In order to survive in the land environment, the invaders also had to develop new sensory organs. The need to find food or to avoid predators would eventually become paramount, although neither was probably of concern to the earliest encroachers. Fish rely heavily on lateral lines for detecting movements in the water, a system that is useless on land. Instead, the senses of sight and hearing had to be adapted to a dry environment. Crossopterygians had eyes, of course, but they required modification to work well on land. The evolution of the eardrum, from a bone that braces the jaw of crossopterygians, has been documented in early amphibians.

Perhaps the most challenging task was to develop a means of reproduction out of water. The sea provides a medium for broadcasting sex cells, in effect seeding the area with offspring. The amphibians never mastered the art of reproducing on land, but reptiles invented the amniote egg, a self-contained thimble of seawater to protect and nourish the developing embryo. This was arguably the most important step of all in the conquest of the land, in that it freed vertebrates completely from their aquatic origins.

The crossopterygians squirmed onto continents that were already greening. The earliest land pioneers were probably algae, which appeared as early as 440 million years ago. The ancestors of liverworts, mosses, and ferns were established as long as 400 million years ago, and forests of towering horsetails populated the land soon thereafter. Plants, too, faced great difficulties in colonizing this new frontier. Like animals, plants on land required more structural support. This problem was solved with the invention of lignin, woody tissue that could support an erect plant. By adding a waxy outer coating (the cuticle, the very signature of the modern land plant), they were able to regulate water loss. Pigments served to minimize the effects of ultraviolet radiation. The earliest land plants remained dependent on water to transport the sexual phases of their life cycles, but soon found other means to spread their spores.

Two basic facts may explain why plants flourished out of the water. First, water attenuates sunlight, which all plants need to thrive. And second, the need for oxygen and carbon dioxide, the raw materials for plant metabolism, is better met on land than in water. But the fossil record of animals gives no obvious clues as to why they may have left the Devonian sea. Perhaps land animals benefited from reduced competition or found the plants there to be nourishing and delicious.

The first land vertebrates, the earliest amphibians, were really just slightly modified fish. Their skulls were almost identical to those of crossopterygians, and they still spent part of their lives in the water, breathing through gills. Their limbs were not drastically different from the paddle-shaped fins of their fishy relatives. The leap to land did not cover a great distance anatomically, but it was one of the most profound in all of evolutionary history. The lesson in this story is the revelation of one of life's strangest qualities—its eternal dissatisfaction with the status quo, its insistence on occupying new habitats by adapting to the most fantastic of circumstances. Perhaps life has no choice but to be flexible and opportunistic, because its environment is always changing.

THE DINOSAURS' DEMISE

Mass extinctions are the benchmarks of the geological time scale, and they have invited speculation for more than a century. The most famous example occurred 65 million years ago at the end of the Mesozoic era, that is, the boundary between the Cretaceous and Tertiary periods (following geologic tradition, we will henceforth call this the "K/T" extinction). This extinction, though not the worst carnage in the geologic record, is certainly the most notorious, since it marks the ultimate demise of dinosaurs. Some paleontologists have pointed to the fact that dinosaurs were already on the decline by this time, but it remains a fact that none of these monsters has been found conclusively on our side of the K/T boundary. The fossil record of dinosaurs is meager, though, so it is difficult to pinpoint the exact time of their disappearance. In contrast, the sudden decimation of many marine animals and plants at the

A thin layer of clay (behind the coin), exposed near the town of Gubbio, Italy, delineates the boundary between Cretaceous strata below and Tertiary strata above. It also marks one of the most profound mass extinctions in the geologic record. High concentrations of the element iridium in this clay indicate the occurrence of a massive meteor impact, and suggest a possible cause for the dying.

K/T boundary is glaringly obvious. Half of the genera living in the oceans simply disappeared. Sponges, bivalves, snails, echinoids, fish, and most types of plankton were severely affected.

In 1978, Luis and Walter Alvarez (father and son) and several of their colleagues at the University of California, Berkeley, made a serendipitous discovery: a thin layer of undistinguished brown clay, sandwiched between Mesozoic and Cenozoic limestones at Gubbio, Italy, contained a remarkably high concentration of the element iridium. They soon documented comparable iridium enrichments in sediments corresponding to the same 65-million-year-old stratigraphic boundary in Denmark and New Zealand. We now know that a similar spike in the abundance of this obscure element, having been found at more than fifty localities, marks the end of the Mesozoic era worldwide.

So what's the big deal? It is this: The abundance of iridium in crustal rocks is very low, normally less than a hundredth of the amount measured in the K/T boundary clay. Iridium is highly soluble in iron metal, and so was presumably scavenged during core formation in the early Earth. Some other elements, such as platinum and gold, share this characteristic affinity for iron, and they too are depleted in crustal rocks and enriched in the boundary clays. Materials that have never experienced core differentiation (chondritic meteorites) contain considerably more iridium than do any terrestrial rocks. The Alvarezes suggested that the worldwide iridium spike must have been caused by a large meteor impact, which spread its matter over the globe.

Following the precepts of uniformitarianism, the K/T extinction had been attributed to gradual processes, such as climate change associated with global reshuffling of the continents. Even gradual processes can lead to sudden biologic crashes, as some combination of cyclic events gets in phase. Instead, the Alvarez team boldly suggested that this wave of death followed a catastrophic impact. The collision of a 10-kilometer-diameter object, they noted, would have lofted huge amounts of pulverized debris into the atmosphere, where it could have circled the globe for years before finally falling back to the ground. The fine particles would have shut out sunlight, halting plant photosynthesis. The demise of many plants would cause a cascade of slaughter in the animal kingdom, the vegetarians starving first and then the carnivores.

Since then, other death-dealing mechanisms related to a large impact have been suggested. As the meteor streaked through the atmosphere it must have heated the air, causing reactions between nitrogen and oxygen to form nitrous oxides (lightning does the same thing, but on a smaller scale). These oxides would fall to the ground as acid rain, and when dissolved in seawater, they could have poisoned marine organisms. On land, heat from the impact may have triggered huge forest fires, with the smoke from this conflagration adding to the already problematic haze. It seems likely that the Earth would have rapidly chilled, due to blocked sunlight and the inability of heat to travel through the dust cloud. A freezing period lasting for weeks or months might kill more plants and add to the discomfort of hungry animals. When photosynthesis was reduced, carbon dioxide accumulated in the atmosphere and in the oceans. More carbon dioxide in seawater would render it acidic, causing the carbonate shells of sea creatures to dissolve.

The timescales for these processes are very different. Environmental destruction within a few hundred kilometers of the impact was probably almost instantaneous. Fires, darkness, and cold temperatures might have lasted for a few weeks or months. The effects of acid rain and toxic substances may have persisted for years or decades. Different species may have been driven to extinction by different causes, allowing some spread in the actual times of their demise.

Understandably, the Alvarez hypothesis was followed by a storm of scientific controversy and a great deal of public interest. Based on physical evidence collected since its proposal, the idea seems to hold up well. Shocked quartz grains, identical to those found in impact craters, have been discovered in boundary clays all over the world. At some locations tiny bits of soot, suggesting burning in the aftermath of the impact, were uncovered. Other suggested mechanisms to explain the iridium spike in boundary clay, such as effusion from volcanoes, appear to be inadequate. But the smoking gun, a large impact crater of just the right age, was nowhere to be found. Of course, an impact into the ocean 65 million years ago would have made a scar on the seafloor that, by now, would have been subducted.

Then, just a few years ago, a gargantuan crater was found in Mexico, right off the tip of the Yucatán Peninsula. The 300-kilometer-diameter pit, named Chicxulub (translated from the Mayan language as "tail of the devil") after a village near Cancún, lies buried under a kilometer of sediments and is half submerged beneath the waters of the Gulf of Mexico. The crater was discovered by petroleum geologists using geophysical data in their search for subsurface riches. More recent geophysical surveys and drilling have confirmed its impact origin. The Chicxulub meteor excavated the largest crater on Earth and simultaneously melted 20,000 cubic kilometers of rock in just a few seconds. Droplets of this melt were splashed outward for thousands of kilometers, to be incorporated into K/T boundary clays in Haiti and the western United States.

There is now little, if any, doubt that the end of the Mesozoic era was punctuated by a massive impact. Many of those who study the fossil record, though, resist the idea that the impact caused the K/T extinction. The skeptics demand proof that species were wiped out instantaneously, and their data seem to suggest that some lingered for thousands or millions of years. Moreover, some species, especially land dwellers like turtles and lizards, survived the impact event unscathed. Such difficulties have prompted some geologists to grapple with more complex scenarios. Perhaps the end of the Mesozoic was already a time of dying, an extended period of environmental stress and ecological deterioration, so that an impact was merely a final blow. Another interesting suggestion is that the K/T extinctions were stepped, prompted by several, closely spaced impacts.

I freely confess that I am a convert to the idea that an impact caused the K/T extinction. It seems improbable, at least to me, that there is no connection between these events, and inescapable that large impacts would do harm to the biosphere. The lesson from the great snuffing of Mesozoic life may be that we cannot ignore catastrophe as a normal geologic process. An ample cratering record documents that many such impacts have happened throughout geologic time, events that, for the most part, have been ignored by those who study biologic extinctions. Environmental stresses that are the causative agents for biologic evolution can sometimes happen in the blink of an eye.

THE EVE OF HUMANKIND

The sparsest definition for the genus *Homo* is an anthropoid mammal that walks upright on its hind limbs and possesses a cranial capacity larger than about 800 cubic centimeters. We are *Homo sapiens* (the wise man), the only living species of this genus. In the past, however, humankind has included other species, as well as another genus apart from *Homo*. Members of this broader family are called *hominids,* whereas an even broader group that also includes apes would be called *hominoids*.

When Darwin published *Origin of Species,* virtually nothing was known about the hominids. Even today, despite some considerable success in collecting man's fossilized predecessors, the record remains scanty. Before the practice of human burial, the chance of preservation was remote. Anthropological bonanzas were found in a few locations where hominids practiced cannibalism or lived in the same cave for many generations, but for the most part human paleontology required luck and a great deal of digging. Early on, every bone fragment was commonly assigned a specific name, exacerbating the problem of classification. This was a tricky enough problem to begin with, because we represent the end of this lineage and it is difficult to be objective about our own ancestry.

The geologic record of human evolution consists almost exclusively of bones and teeth, inherently exciting I suppose to orthopedists and dentists, but not the sort of stuff by which we ourselves recognize each other. Early classifications based on anatomy lumped apes into a distinct group from all the past and present members of the genus *Homo*. Since the 1960s, however, molecular biologists have waded in with their discovery that, at the genetic level, chimpanzees and gorillas are much closer kin to humans than their anatomy might suggest. The orangutans are less closely related and presumably branched off earlier from the line that led to hominids.

A crucial piece of mankind's evolutionary puzzle appeared in 1974, when Donald Johanson, working in the Afar region of Ethiopia, found the fossilized remains of a primate. The hominid stood only three and a half feet tall, but walked upright. While Johanson examined the skeleton in his tent, the Beatles' "Lucy in the Sky with Diamonds" played on the radio. Thinking this small creature was female, Johanson called her "Lucy," a less formidable name than the one he gave to her species, *Australopithecus afarensis*. Lucy and her kin first appeared about 4 million years ago. Subsequent discoveries suggest that the genus *Australopithecus* contained other species at different times and possibly at different locations. *Australopithecus africanus* may be somewhat younger, and *Australopithecus ramidus* (from the word for "root" in the Afar language) is thought to be somewhat older. The actual evolutionary relationships between these species are mired in controversy. Let us simply say that these creatures tread the line between hominids and apes, and may have appeared sometime after about 5 million years ago when gorillas and chimpanzees diverged. Collectively, they are termed *gracile Australopithecines,* because they were small and slender.

Perhaps 2.5 million years ago, the gracile Australopithecines branched confusingly into three separate but coexisting species. Two of these, *Australopithecus boisei* and *Australopithecus robustus,* led nowhere. These large, brutish hominids, which soon perished, were anatomically distinct from their neighbor, *Homo habilis* (the handy man).

In 1972, Richard Leakey found the first fossils of the genus *Homo habilis* in Kenya. All of the various species of *Australopithecus* had brains that were slightly smaller than the modern gorilla, whereas *Homo habilis* had a brain half again as large, despite his pygmylike size. This creature was a toolmaker and probably ate meat as well as seeds, roots, and bugs. It is possible that *Homo habilis* also had a rudimentary form of speech, as inferred from an enlargement of the portion of the brain that controls language.

Homo habilis was the direct ancestor of *Homo erectus* (the upright man), the next stage of evolution of the hominids. Fossil remains of this species, some as old as 1.5 million years, have been found in many parts of the world, as can be discerned from the names given some of these speci-

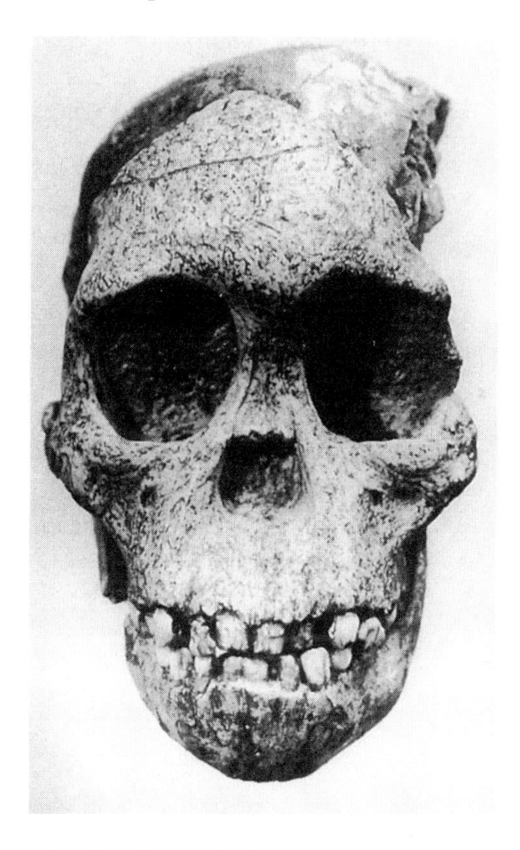

The face of man-to-be. The Taung skull, the type specimen of *Australopithecus*, was discovered in 1924 in a limestone quarry in South Africa. Based on its teeth, the creature is thought to have been a young child.

mens: Java man, Peking man, and Heidelberg man. Java man was actually the first hominid fossil ever recognized, dating from the end of the nineteenth century. As a first approximation, *Homo erectus* was probably indistinguishable from a modern human, but his skull had a prominent ridge and his jaw showed no trace of a chin. The brain capacity of *Homo erectus* was about equal to that of a modern four-year-old child. He made a variety of flint tools, learned to control fire, and eventually traveled the world.

Our own species, *Homo sapiens,* evolved from *Homo erectus,* and the fossil record contains many transitional forms between the two. Fossils with intermediate features have compounded the difficulty in deciding exactly when *Homo sapiens* appeared, but a reasonable estimate is about half a million years ago. Early varieties, such as the Neanderthals, were large-brained but still retained some archaic anatomical features of their

Hominid Evolution

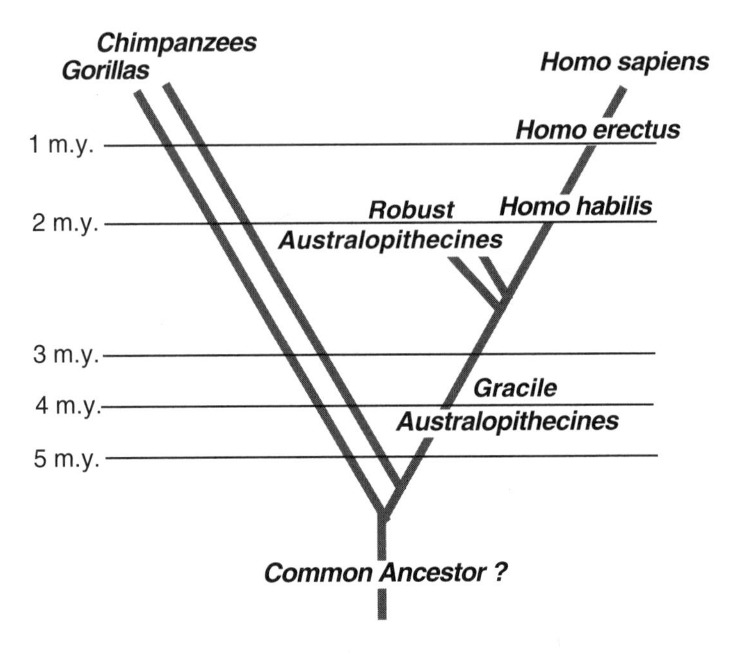

Principal divergences in the evolution of hominids are illustrated by branches in this family tree, along with the approximate times of speciation events (in millions of years). The various species of *Australopithecus* are lumped together into gracile and robust groups.

erectus ancestry. The modern version of *Homo sapiens* probably dates from no more than several hundred thousand years ago.

Despite the fact that the fossil record of hominids is based only on bones and teeth, there are many features that can be used to differentiate the various species. Some features that characterize modern humans appeared early, and some late. For example, brain size evolved slowly in early forms, attaining most of its increase within the last half million years. Conversely, the evolution of hands and teeth reached almost modern form in *Homo habilis* earlier than 1 million years ago. Such a pattern of change, with different characters evolving at different rates, is called *mosaic evolution*. It also appears that early hominids, like modern apes, matured to adulthood at considerably faster rates than does modern *Homo sapiens*. Prolonged development has had important consequences, not the least of which is the size of our brains. In living primates other than humans, brain growth slows at birth; however, in humans it continues at a rapid pace for about a year beyond birth, accounting for its greater size. (It has been suggested that humans actually have a twenty-one-month gestation period, with the last twelve months outside the womb. Without such "premature" arrival, the birth of babies with large brains would pose an impossible hurdle for mothers.) Changes in developmental timing, processes collectively known as *heterochrony,* may have allowed major alterations, and perhaps even new species, to arise within only a few generations.

In addition to the morphological fossil record, the evolution of *Homo sapiens* has been traced, interestingly enough, through its biochemistry. A little background is necessary here. As in all organisms, genetic material in humans (the DNA in cell nuclei) is contributed by both parents. Other parts of cells, however, may contain genetic matter derived from only one parent. For example, the DNA blueprints for mitochondria (abbreviated mtDNA), which carry out respiration in the cell, are transmitted by the mother alone—the father's mtDNA is discarded. Consequently, all human mitochondrial DNA resembles that of the original female parent from which it was first derived. This mtDNA does not affect our appearance, as mitochondria only control respiration.

In 1987, Rebecca Cann and her colleagues at the University of California traced lineages of mtDNA in living people. By employing a molecular clock based on the rate of accumulation of small mutations in mtDNA, they were able to construct family trees. The limbs of these trees converged on a trunk, a single source for human mtDNA who lived about 150,000 years ago. This source could be a small number of females, or perhaps even one woman. It was probably inevitable that this source would be dubbed "Eve," or more accurately "mitochondrial Eve." (A cautionary note: The originator of our mtDNA was not the only woman in her breeding population; the mtDNA from other maternal lines with no daughters eventually disappeared, leaving only one source.) From differences in mtDNA among various peoples around the world, we can infer that Eve lived in Africa or Asia. The fossil record gives a consistent picture, with the oldest specimens of modern *Homo sapiens* found in rocks several hundred thousand years old on those two continents. This work remains controversial, however, because mtDNA only tells the story of one gene's heritage.

The lesson, plain and simple, from hominid evolution is that the modern human clearly was not an inevitable design for an intelligent being. In the transition from ape to man, there were false starts and dead ends. We are the current product of an evolutionary path derived from

This cartoon is based on the common perception that evolution is progressive. The sign, however, hints at the fact that nothing is sure in evolution. (*Frank and Ernest,* reprinted by permission of Newspaper Enterprise Association, Inc.)

the genus *Australopithecus* and winding through other species of the genus *Homo,* the lone survivor from several rival lineages. The characteristics we think of as human were appended to hominids at various points in their evolutionary history, and more than a little of our advancement can be attributed to prolonged developmental periods in later species. At any point in the struggle for survival, hominid evolution could have taken a different turn, so that this chapter would have had a very different ending.

THAT'S PROGRESS

Certainly, Darwin could have portrayed his theory of evolution in a less threatening way. It would perhaps have been easier to swallow if cast in the context of *progressive* change, an orderly ladder leading sequentially to higher life forms with mankind at its pinnacle. Many evolutionists, before and after Darwin, have portrayed biologic change in just that way. As a matter of fact, the very word *evolution* derives from an English vernacular usage meaning "progress." Darwin did not like the term "evolution," preferring the more cumbersome phrase "descent with modification," but on this semantic point he lost.

The cauldron of evolution demands only change, not necessarily progression. In viewing the sweep of life's history, it is difficult not to view today's complex fauna as being a step or two above their primitive ancestors. But progress up the evolutionary ladder, if that's what it is, was not linear. Instead, change happened in fits and new starts, by trial and mostly error. It has never been a pretty picture, this survival of the fittest, or sometimes of the luckiest. But its results, even if undirected, are grand.

We should understand that evolution is not just history, although that is how we teach it. At the same time that we have come to appreciate past life's innumerable varieties and to comprehend its relentless drive for modification, we halt suddenly at the present day. We humans often visualize ourselves as no longer changing, as a culmination, even when we try to project into the future. That is a mistake. The work of evolution is not finished, even for us. I like to think humankind may improve in time, though that certainly is not a given. Another expressed this thought more elegantly:

> We look back through countless millions of years and see the great will to live struggling out of intertidal slime, struggling from shape to shape and from power to power, crawling and then walking confidently on the land, struggling generation after generation to master the air, creeping down into the darkness of the deep; we see it turn upon itself in rage and hunger and reshape itself anew, we watch it draw nearer and more akin to us, expanding and elaborating itself, pursuing its relentless inconceivable purposes, until at last it reaches us and its being beats through our brains and arteries. . . . It is possible to believe that all the past is but the beginning of a beginning and all that has been is but the twilight of the dawn. It is possible to believe all that the human mind has accomplished is but the dream before the awakening. . . . Out of our . . . lineage, minds will spring, that will reach back to us in our littleness to know us better than we know ourselves. A day will come, one day in the unending succession of days when beings, beings who are now latent in our thoughts and hidden in our loins, shall stand upon this earth as one stands upon a footstool, and shall laugh and reach out their hands amidst the stars.
>
> —H. G. Wells, "The Discovery of the Future"

Some Suggested Readings

Briggs, D. E. K., Erwin, D. H., and Collier, F. J. (1995). *The Fossils of the Burgess Shale*. Smithsonian Institution Press, Washington. *This collection of exquisite protographs and drawings of the Smithsonian's collection of Burgess*

Shale fauna complements the story of their discovery and study by Gould (see below).

Desmond, Andrian, and Moore, James (1991). *Darwin*. Warner, New York. *This thick book is an enjoyable foray into the life and times of Charles Darwin, providing rich insights into the evolution of Darwin's theory and its reception; a great read.*

Glen, William (1990). What killed the dinosaurs? *American Scientist,* vol. 78, pp. 354–70. *A nice review of the controversy about the cause of the K/T extinction; this paper tries too hard to be even-handed for my taste, but the reader will benefit from exposure to alternative ideas.*

Gould, S. J. (1989). *Wonderful Life: The Burgess Shale and the Nature of History*. Norton, New York. *A most elegant and insightful summary of the discovery of the Burgess Shale and the scientific study of its miraculous fauna.*

Long, J. A. (1995). *The Rise of Fishes*. Johns Hopkins University Press, Baltimore. *A gloriously illustrated book all about the evolution of fish; crossopterygians and their leap onto the land are described in detail.*

McGowan, Chris (1984). *In the Beginning*. Prometheus, New York. *A scientist shows why the creationists are wrong; Chapter 14 provides an easily grasped summary of the evolution of hominids and the fossil discoveries that led to this understanding.*

Silver, L. T., and Schultz, P. H., eds. (1982). Geological Implications of Impacts of Large Asteroids and Comets on the Earth. *Geological Society of America Special Paper,* 190. *A thorough, though now slightly dated, discussion of the controversy of relating extinctions to impact events.*

Thomson, K. S. (1991). *Living Fossil: The Story of the Coelacanth*. Norton, New York. *This fascinating and easily understood book gives an unabridged version of the discovery and scientific study of the coelacanths.*

Tobias, P. V. (1992). Major events in the history of mankind. In *Major Events in the History of Life,* edited by J. W. Schopf, Jones and Barlett Publishers, Boston, pp. 141–75. *A masterful synopsis of hominid evolution, including interesting sections on ecology and culture.*

Epilogue

Pierre Teilhard de Chardin (1881–1955) was an accomplished scientist and philosopher, as well as a Jesuit priest. At the age of forty-one, he earned a doctorate in paleontology from the Sorbonne, which seemed to presage a brilliant university career. However, Teilhard's unconventional philosophical ideas and his uncanny ability to articulate them before large audiences soon got him into hot water with his ecclesiastical superiors, who effectively "exiled" him from his native France for the bulk of his scientific career. Most of this time was spent in China, where Teilhard was part of an international scientific team that discovered the fossil remains of Peking man. In due course, his exile took him to every important site of prehistoric human life known in his day. Teilhard sought to explain the relationship between his scientific studies and his religious beliefs, but his order prohibited him from publishing anything other than purely scientific works. His fame as a natural philosopher came only after his death, when his friends posthumously published his other writings.

Teilhard was a contemporary of Sir James Jeans (the British mathematician whose harsh view of the inconsequential place of the Earth in the cosmos introduced this book), and the pair provide an interesting contrast. Jeans was phenomenally successful and garnered most of the accolades that his profession and the broader public had to offer; Teilhard was largely ignored in his time, except by his Jesuit superiors, who muzzled him to keep him quiet. Jeans looked upward to the stars and planets as the scientific basis for his philosophy; Teilhard's attention was mostly directed toward the fossil record below his feet. Jeans was awed by the sheer immensity of space and time, which led him to a rather gloomy assessment of earthly and human conditions; Teilhard's perspective on the importance of the Earth and its inhabitants was hopeful and triumphant, and so of necessity, critical of Jeans's philosophy:

> This [*Jeans's*] bleak vista is not only so discouraging as to make action impossible; it is so much at variance, physically, with the existence and exercise of our intelligence that it cannot be the last word of science. Following the physicists and astronomers we have thus far been contemplating the Universe in terms of the Immense. . . . But is it not possible that we have been looking through the wrong end of the telescope, or seeing things in the wrong light?
>
> —Pierre Teilhard de Chardin, *The Future of Man*

It is with a brief restatement of one modest aspect of Teilhard's soaring natural philosophy that I choose to end this book. I do not wish necessarily to champion the bulk of Teilhard's unconventional ideas, but his view of the importance of the Earth and its life is relevant and thought-provoking.

From what we have learned from astronomy, planets like the one on which we live certainly do appear to be an insignificant part of the universe. The uncountable stars, including the nearest one in whose glare we bask, seem gargantuan by any measure, whether mass, energy, or longevity. These mighty engines of the night are clumped together into galaxies of almost unfathomable size and number. And this is just the observable part of the universe. Interstellar dust, gas, and the cinders of stars, the dark, invisible matter nestled in the vast spaces between the stars, may actually comprise most of the mass of the cosmos. Astronomical distances are, well, astronomical, logged in light years and parsecs. The 4.5-billion-year age of our own solar system boggles the mind, and we now know that our metal-rich Sun formed from the ashes of even more ancient stars in conflagration. It is easy, as did Jeans, to feel lost in such a universe.

Far out between two of its spiral arms, the Milky Way nurses a modest yellow star, our Sun. This particular location is unusual, in that a star here takes an especially long time to cross into the next spiral arm and thus is relatively safe from supernova explosions that are common within the arms. Moreover, the star's oscillation period out of the midplane of the galaxy is anomalously low, so that it is shielded by gas and dust from harmful radiation emanating from the galactic center. Our star is the right generation for its stellar ancestors to have forged enough metals to make rocky planets and the right size to be energetically stable for billions of years.

The third planet outward from this star, a tiny blue sphere, is hardly conspicuous, until it is examined closely. This planet is mostly inundated with liquid water, due to a collusion of forces—the fortunate happenstance of its distance from the Sun and its own mass, the partial degassing of its interior and perhaps the late addition of a veneer of comets. The stabilization of its orbital obliquity by a lone but rather massive satellite (itself the product of a huge impact early in its history) has damped its climate fluctuations. Sometime after 4 billion years ago carbon-based life forms appeared in the water on this world, and they have been spreading and evolving ever since. Teeming green organisms, incapable of thought or conscious action, now form a synergy with the planet, having altered the very composition of its atmosphere to allow other, more complex organisms to have emerged from the seas. The land dwellers thrive on jutting platforms of differentiated siliceous rock that move restlessly about the surface, a unique topology made possible by its internal heat engine. A terrestrial organism capable of self-conscious reasoning appeared very recently and almost immediately dominated its ecosystem. Even when we consider only the history of the Earth itself, some humility is in order for our species.

Teilhard's answer to the seeming unimportance of our tiny world and the self-cognizant animals that cling to its surface was to redefine the universe in terms of *complexity,* as an alternative to quantity. He viewed complexity not as a matter of simple multiplicity of constituents but of organization, which takes into account the number and variety of linkages between the constituents. Viewed through the other end of Teilhard's telescope, an atom is more complex than its constituent nuclear particles and electrons, a molecule or a mineral grain has higher complexity than its constituent atoms, a cell is yet more complex, and a living organism possesses an even higher order of complexity. It is not galaxies, stars, or planets that are the key units of Teilhard's reality because they are simply very large aggregates lacking complex organization. So it is not hard to see where he put *Homo sapiens* in this hierarchy. Teilhard's scheme obviously has metaphysical and religious overtones, and its scientific basis has been both criticized and supported by a number of prominent scientists. In its defense as a scientific classification (science certainly permits more than one valid way of classifying things), I will note that, when arranged according to Teilhard's scale of complexity, things succeed one another in the same order of their first appearance in nature. Teilhard himself suggested that classifying things in this way provides a bridge between the troublesome and seemingly irreducible gaps separating astronomy, physics, chemistry, geology, and biology.

With Teilhard's complexity scale in mind, let's look again at the universe and the Earth's place in it. Huge, incandescent stars unleash torrents of energy, but they are, after all, only factories in which nature, starting with hydrogen atoms, manufactures heavier elements. Stars are nuclear reactors that fabricate matter occupying the lowest rung on the ladder of complexity; they can do no more. The vast interstellar medium collects these elements and, over time, synthesizes some portion of them into simple molecules or mineral grains, thereby ascending to the next level. But reaching higher rungs of complexity, that is, constructing large organic molecules and thence life, can happen only in a few, seemingly insignificant places, a relative handful of favorably disposed worlds, perhaps separated by such vast distances that they will never know of their common existence. As described more eloquently by Teilhard:

> Despite their vastness and splendour the stars cannot carry the evolution of matter much beyond the atomic series: it is only on the very humble planets, on them alone, that the mysterious ascent of the world into the sphere of higher complexity has a chance to take place. However inconsiderable they may be in the history of sidereal bodies, however accidental their coming into existence, the planets are finally nothing less than the key-points of the Universe. It is through them that the axis of life now passes; it is upon them that the energies of an evolution principally concerned with the building of large molecules is now concentrated.
>
> —Pierre Teilhard de Chardin, *The Future of Man*

We do not yet understand the steps leading from the building of large molecules to the appearance of living organisms. Our own solar system tells us, however, that the conditions where life appears and thrives must be special, even for planets. It is astonishing that there is so much universe and apparently so little life, but the building of organized complexity is grossly inefficient. The Earth and its inhabitants may exist because of a tenuous connection of seemingly chance events amidst nature's prodigious waste of energy and substance. But its success, however improbable, is something to be celebrated, not be belittled. Teilhard trumpeted these feelings, in what might accurately be described as a fanfare for Earth:

> We can only bow before this universal law whereby, so strangely to our minds, the play of large numbers is mingled and confounded with a final purpose. Without being overawed by the improbable, let us now concentrate our attention on the planet we call Earth. Enveloped in the blue mist of oxygen which its life breathes, it floats at exactly the right distance from the sun to enable the higher chemisms to take place on its surface. We do well to look at it with emotion. Tiny and isolated though it is, it bears clinging to its flanks the destiny and future of the Universe.
>
> —Pierre Teilhard de Chardin, *The Future of Man*

Teilhard's statement notwithstanding, science does not demand or even accept that our species is the pinnacle or final purpose of the universe. Yet who cannot be awed by the unlikely existence of consciousness, and by the fragile blue planet that cradles it? Not awed into intellectual stupor or blind acceptance of creed, but rather inspired anew to unravel the mysteries of our unique world's origin and evolution. As expressed by the poet:

> *We shall not cease from exploration*
> *And the end of all our exploring*
> *Will be to arrive where we started*
> *And know the place for the first time.*
>
> —T. S. Eliot, *Four Quartets*

Sources of Quotations

Dedication

Sondheim, Stephen, "Children Will Listen" (from the musical *Into the Woods*). © 1988 Rilting Music, Inc. All rights administered by WB Music Corp., all right reserved. Used by permission of Warner Bros. Publications U.S. Inc.

Prologue

Cronin, Vincent (1981). *The View from Planet Earth.* Morrow, New York.

Eiseley, Loren (1964). *The Unexpected Universe.* Harcourt Brace Jovanovich, San Diego.

Jeans, James (1931). *The Mysterious Universe.* Macmillan, New York.

Milne, E. A. (1952). *Sir James Jeans, A Biography.* Cambridge University Press, London.

Genesis to Geology

Keynes, J. M. (1946). *Newton Tercentenary Celebrations.* Read before The Royal Society and published (1951) in *Essays in Biography,* Norton, New York.

Stardust and Antique Elements

Gamow, George (1952). *The Creation of the Universe.* Viking Press, New York.

Lemaître, Georges (1951). *The Primeval Atom.* Van Nostrand, New York.

Simmer Until Done

Goldschmidt, V. M. (1922). Der Stoffwechsel der Erde. *Vidensk. Skrifter. Math.-naturv. klasse,* no. 11, 25 p.

Rock of Ages

Eve, A. S. (1939). *Rutherford.* Macmillan, New York.

Huxley, T. H. (1868). A liberal education; and where to find it. *Collected Essays,* vol. 3, Macmillan, London.

Twain, Mark (1962). *Letters from the Earth.* Bernard DeVoto, editor. Harper, New York.

Terra Not Firma

Saint-Exupéry, Antoine de (1943). *The Little Prince.* Translated from the French by Katherine Woods. Hartcourt, Brace & World, New York.

High-Water Mark

Margulis, Lynn, and West, Oona (1993). Gaia and the colonization of Mars. *GSA Today,* vol. 3, no. 11, pp. 277–91.

Let There Be Slime

de Duve, Christian (1995). The beginnings of life on Earth. *American Scientist,* vol. 83, no. 5, pp. 428–37.

Eiseley, Loren (1957). *The Immense Journey.* Random House, New York.

Monkey Business at the Seaside

Eiseley, Loren (1957). *The Immense Journey.* Random House, New York.

Gould, S. J. (1994). The power of this view of life. *Natural History,* vol. 103, no. 6, pp. 6–8.

Wells, H. G. (1902). The discovery of the future. *Nature,* vol. 65, no. 1684, pp. 326–31.

Epilogue

Eliot, T. S. (1943). "Little Gidding," in *Four Quartets.* Copyright 1943 by T. S. Eliot and renewed in 1971 by Esme Valerie Eliot, reprinted by permission of Harcourt Brace & Company.

Teilhard de Chardin, Pierre (1964). *The Future of Man.* Translated from the French by Norman Denny. Harper & Row, New York.

Index

About the Author

Harry Y. McSween, Jr., is a professor and head of the Department of Geological Sciences at the University of Tennessee. After graduating from The Citadel (B.S. chemistry, 1967) and the University of Georgia (M.S. geology, 1969), he served as an Air Force pilot during the Vietnam era. In 1977, he received a Ph.D. in geology from Harvard University. Since that time, McSween has been a member of the faculty of the University of Tennessee and a NASA principal investigator, with research interests in meteorites and the geology of asteroids and planets, especially Mars. He has served on numerous NASA and National Academy of Sciences advisory panels dealing with space exploration, and has been a science team member for several spacecraft missions. He is also past president of the Meteoritical Society, an international organization devoted to the study of meteorites and planetary science. Dr. McSween's previously published books are *Meteorites and Their Parent Planets* (Cambridge University Press, 1987), *Geochemistry: Pathways and Processes* (Prentice-Hall, 1989, coauthored with S. M. Richardson), and *Stardust to Planets: A Geological Tour of the Solar System* (St. Martin's Press, 1993; republished in softcover, 1995). He lives with his family in Knoxville.